MATTEO IANNEO

MESSAGES FROM MARS

I

Introduction

Mars is the fourth planet from the Sun and is known as the "Red Planet" because of the enormous amounts of iron on the surface. Its dimensions are approximately half those of the Earth. It has two moons: Phobos e Deimos. Mars orbits the Sun at an average distance of about 228 million kilometres and its orbital period is about 687 days, corresponding to 320 days, 18.2 hours on Earth. The solar day on Mars is slightly longer than ours: 24 hours, 39 minutes and 35.244 seconds. Its atmosphere is mainly composed of carbon dioxide. It was explored for the first time in 1965 by the Mariner 4 space probe which transmitted the first images of the Red Planet, while in 1971 Mariner 9 produced a complete map. Then, in 1976, two further probes were sent, Viking 1 and Viking 2 which, landing on the surface, were able to obtain data about the atmosphere and the temperature on this planet, as well as send images. The light regions in the south are covered by frozen carbon dioxide. Its diameter is about 6,794 km and temperatures are from around -130° to +27° C. Both the Martian polar caps consist mainly of ice covered by a layer of about 1 metre of frozen carbon dioxide, dry ice, at the North Pole, while the same layer reaches 8 metres at the South Pole. In 1609 Galileo Galilei was the first to point his telescope at the Red Planet. After more than two centuries, precisely in 1877, Giovanni Schiapparelli in Milan used a 22cm telescope to make the first detailed map of Mars and its canals. This planet has many fascinating mysteries, including the canals present on its surface, which seem to be river beds surrounded by vegetation, and the various images called pareidolia, created by the action of the strong Martian winds. In Roman mythology, Mars, known as Ares by the Greeks, was considered the god of war, thunder, rain, nature and fertility. The Babylonians called it Nergal, the god of war, fire and destruction, because of its red colour. In Hindu mythology it has the Mangala, while the Jews called it Ma'adim and the Arabs, al-Mirrikh. So, this divinity is found in various mythologies in human history. In the past it was thought that Mars was populated by Martians. In recent years of exploration, thanks to advanced technology, the conclusion has been reached that Mars was not inhabited by civilisations which could have left their mark on this wonderful planet. In other words, Mars is an uninhabitable planet but we don't know definitively whether it has been populated by living beings in the past, but this is what scholars confirm at this time.

Preface

I've always thought in a different way. I have never accepted what others have tried to affirm and to dictate via messages, videos, magazines and so on. I've always believed that the truth, hidden "by the greats of the Earth", has been manipulated to direct and redirect our thoughts to where they are wanted, for fear of revealing the true origins of our humanity. Using Google Earth I have started to analyse maps from the space agencies NASA and ESA, spending hours of my time looking for some details that strike my senses. Like this I have discovered several elements that have contributed to strengthening my convictions and I believe that history must be completely rewritten. Observing and processing various elements in detail, my mind associated them with elements already known, because they are seen in history books. In the various places examined, I found sphynxes, depictions of animals (lions, wolves and other beings) pharaohs, temples and other ruins of ancient civilisations. Speaking to authorities to get feedback and explanations, I have always and only received canned answers, i.e. that what I have observed are effects due to shadows or which have been created over time by natural events which have shaped these representations to make them compatible with our own planet. Experts in my country have offended and humiliated me and have said just about everything. I must share with you the following beautiful phrases which impressed me:

1.   "We have carefully examined the material you sent us, and unfortunately we confirm the same line as other more quoted publications, in not giving any importance to your research."

2.   "This is a very well known phenomenon of perception of the human mind: to be able to, or to want to see human faces in everything. This feature of the human brain has been widely studied and has the scientific name of pareidolia."

3.   "The fact that almost no 'serious' magazine has published anything about this should give you pause for thought."

4.   "I'm not sure that this is the evidence that will convince the sceptics."

5.   "In fact... your 'research' is the height of misinformation, that contributes to covering up the REAL researchers of extraterrestrial anomalies."

6.   "However, addressing this topic and continuing to discuss it or give it publicity is totally USELESS."

7.   "If this is your personal hobby and it gives you satisfaction and gratification please continue in this way... there will always be many people who will give you credit... there is such a complete lack of common sense these days!" "It is not only YOU who must provide the PROOF that there is really this kind of 'face', but it would be in your own interest, so that many people (including ALL the media that you contact) remain not only sceptical, but also distance themselves from this kind of 'research'."

My thanks to all those who have written me these responses, because I am a polite person. I have never offended anyone and I have never declared that my results are 100% certain to be artificial constructions, I have always been cautious about everything I have examined, and I only wanted to highlight these details before others did likewise. The fact remains, with my own effort and with no help from anyone, my name has leapt onto the internet and is quoted in several sites around the world, including that of Google Earth which complimented me on the most clicked *tag* for that period. Apart from all this, I now turn to you and invite you to follow me on this long and wonderful voyage...

My first discovery was made in September 2009 at 3am Italian time. I was in front of my pc, with my 56k modem that allowed me to wander, albeit with difficulty, around planet Earth using the Google Earth *tool*. I live in Cerignola, in the province of Foggia in Puglia, Italy. I live in a suburban are of the town where, in the absence of high-speed internet, I have no option but to resort to a simple, but still effective traditional *dial-up* service.

While I was looking at my town from above, I found myself in the suburbs, with my eyes fixed on my own house, easily identifiable in that it is located near to an expanse of olive trees. I immediately thought that is was nice to be able to wander around the world looking at amazing places which are difficult to visit in person, if not just in our imagination. I looked at everything: roads, woods, monuments. I was in Egypt and I scanned the pyramids, I want to France, then to Spain and to New York. In fact, all the places I wanted to visit.

I didn't know how Google Earth worked, and so fiddling here and there, I discovered other functions. Looking at the software menu, I noticed that there was the option to add light, and more. While I was trying out some of the options, I noticed a link which allows you to see the planet Mars thanks to the space agencies NASA and ESA. I couldn't believe my eyes. And so, with some difficulty as a result of the data transmission speed, I managed to load the sphere of the Red Planet and started to explore it in the same way as I had done with planet Earth.

I saw nothing but reddish earth, holes due to the impact of meteorites, rocks and nothing else. I thought that the scientists were absolutely right in asserting that Mars was a planet made of only sand and rock, a planet long since dead. As I was about to turn off the computer to go to bed, I noticed, zooming in on one area, the presence of a shape similar to a shell. As I enlarged that detail, the shape increasingly took the form of a human face. Yes, really a human face, framed in profile, with many anatomical details clearly distinguishable: ear, nose, eye, eyebrow, mouth and neck - a complete face.

In the meantime, I remembered that I had left a pan full of water on the cooker to prepare a chamomile tea. I ran to the kitchen immediately only to find that the water had completely evaporated. I switched off the gas and returned to my pc. As I continued to look at the face, suddenly there was a power cut. I was desperate. I didn't know if the position of the shape I had found had been saved automatically. Unfortunately, it hadn't. My wife, meanwhile, called me to ask what had happened and I told her simply that there had been a power cut. I went to bed, but my thoughts were troubled by the face, the beginning of a series of amazing discoveries.

The next day, while having lunch with my family of three wonderful children and my beautiful wife who I would change for no other, my youngest son, stopped eating to look for his school on Google Earth. A few minutes later he called me to ask for some help with the search. I got up from the table and walked towards him. I explained some technical things to him, even though I realised that because he was young, he couldn't understand all of my explanations intended to facilitate his search. I returned to sit at the table and continued to eat even though I invited the little guy to return to sit with us before his soup got cold.

My son is stubborn and often doesn't listen to our calls to order. After about ten minutes, I heard him ask out loud "How come the Earth is red all over?". I went over to him and realised that the red Earth he was talking about was none other than Mars. "How come" he continued "there are no trees in this part of the Earth, no sea, nothing?" I told him that it wasn't Earth, but the planet Mars and so he asked me what Mars was. I told him that he would know everything at school after a few years. While he was looking around parts of the planet, I noticed an area familiar to me; it was the place I had seen the night before. I immediately told my son to get up and give me his place because I wanted to continue my research.

About ten minutes passed before I suddenly rediscovered the human face. I jumped for joy and begged everyone not to lay a finger on the pc, again fearing that I might lose all trace of my first sensational discovery. Without hesitation I wrote the coordinates of the image on a notebook. It's difficult to accept all of this, psychologically, and to anticipate how this element might be challenged in the world.

I began to send the news to some websites that deal with this kind of subject, but didn't get any response. After more than a year, I continued in this way, sending an email to other websites, which then, in a short time, disseminated the *Face of Gandhi* around the world.

This was my first discovery and the start of my research into the mysteries of Mars. I have become convinced that this planet is hiding another truth: that of a people who lived in times far removed from us, and whose survivors were helped to move to other planets. But this is only my hypothesis. I believe that life on Earth has undergone a genetic influence from other worlds: the presence of "types" of human race on Earth could be the result of people from other places in the Universe having landed here in the past.

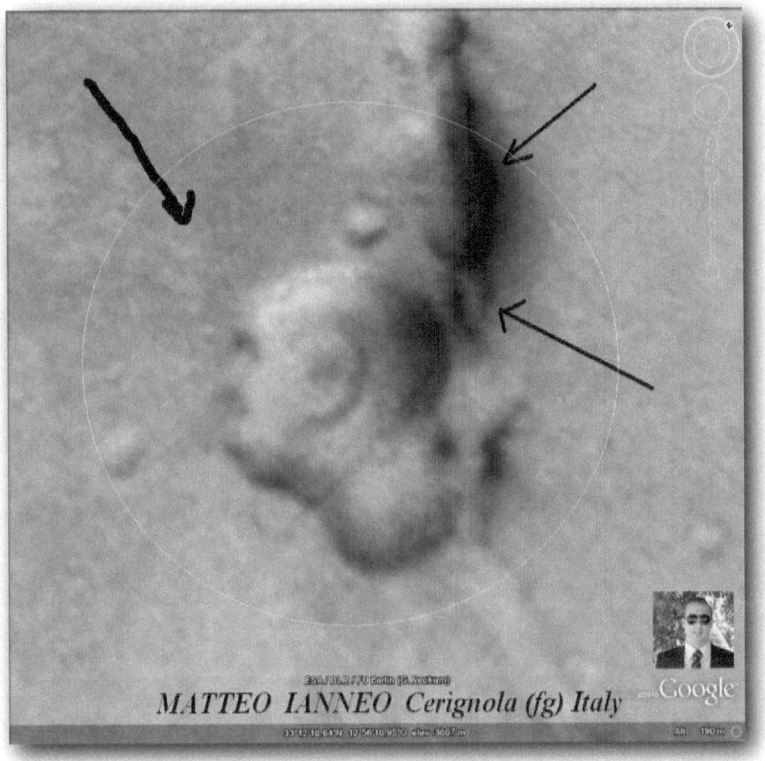

Image source: ESA/DLR/FU Berlin (G.Neukum)

Here is the *Face of Gandhi*. This face is very important to me, many human elements and aspects can be seen: the round head with a dark part that seems to represent the subject's hair, an ear, an eyebrow, an eye, the nose, mouth and neck. I have to say that it really is a good profile. Of course, I leave it to the reader to formulate an opinion and/or personal judgement.

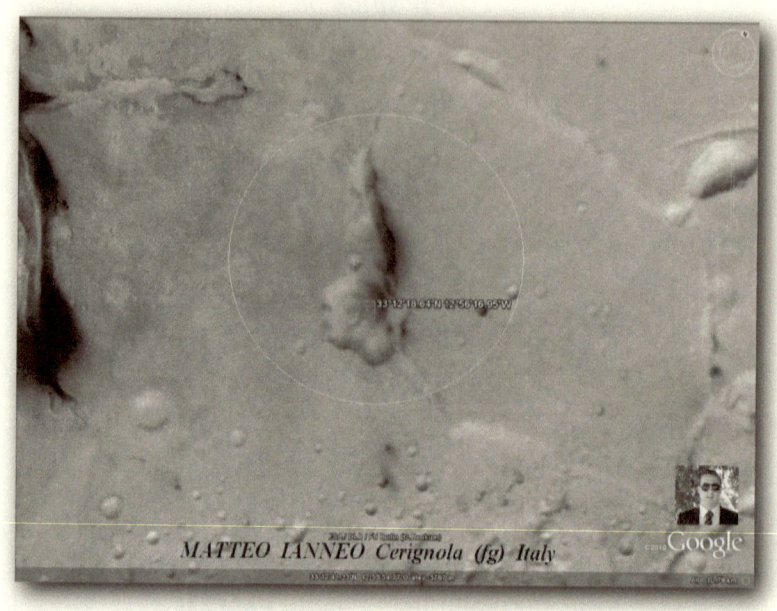

Image source: ESA/DLR/FU Berlin (G.Neukum)

This perspective, seen form a certain elevation, shows the representation of the head with all its details.

Image source: ESA/DLR/FU Berlin (G.Neukum)

In this image a face with a moustache can be seen to the upper left, buried over time. We find ourselves facing the ruins of an ancient civilization.

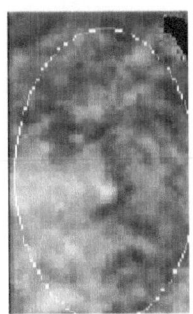

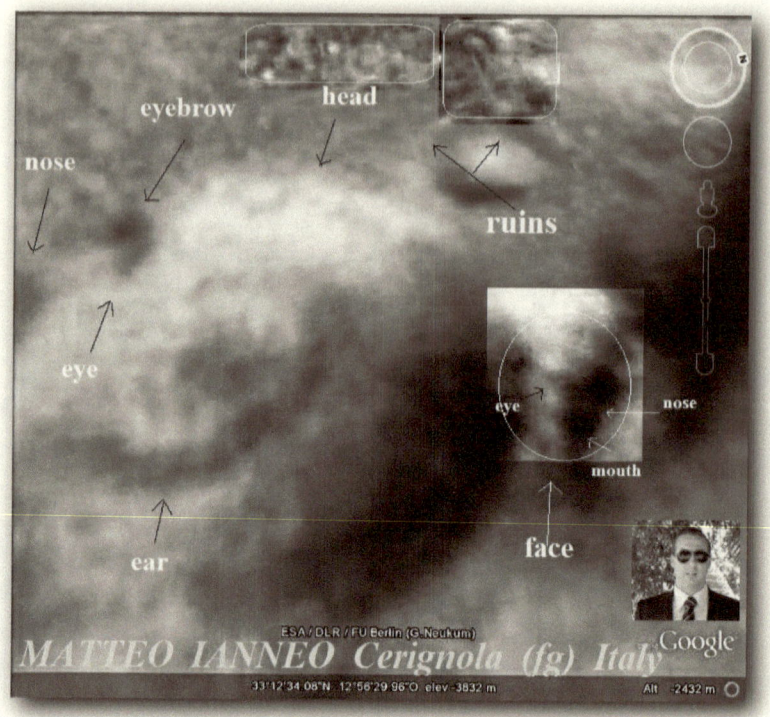

Image source: ESA/DLR/FU Berlin (G.Neukum)

We proceed with a more detailed description of the face: nose, eyebrow, eye, head, forehead,coloured neck (hair), and ear.

Perspective 1

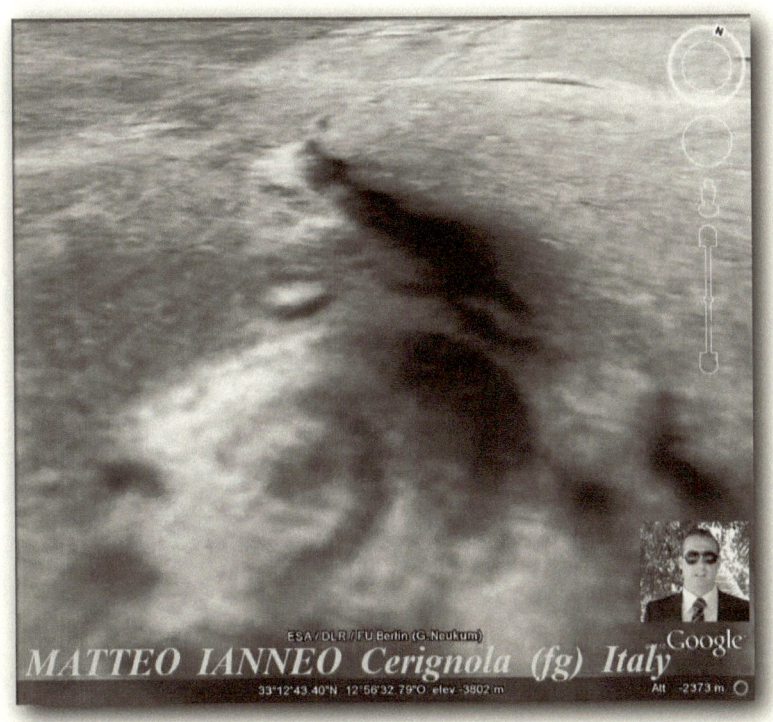

Image source: ESA/DLR/FU Berlin (G.Neukum)

Perspective 2

Image source: ESA/DLR/FU Berlin (G.Neukum)

Perspective 3

Image source: ESA/DLR/FU Berlin (G.Neukum)

Perspective 4

Image source: ESA/DLR/FU Berlin (G.Neukum)

Perspective 5

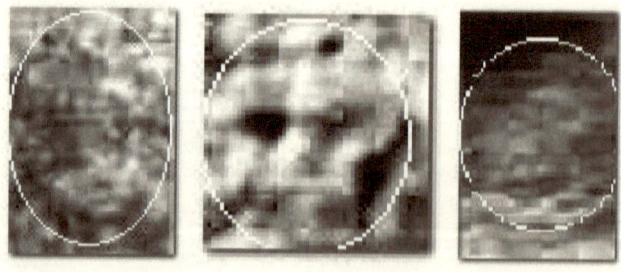

Image source: ESA/DLR/FU Berlin (G.Neukum)

Image source: ESA/DLR/FU Berlin (G.Neukum)

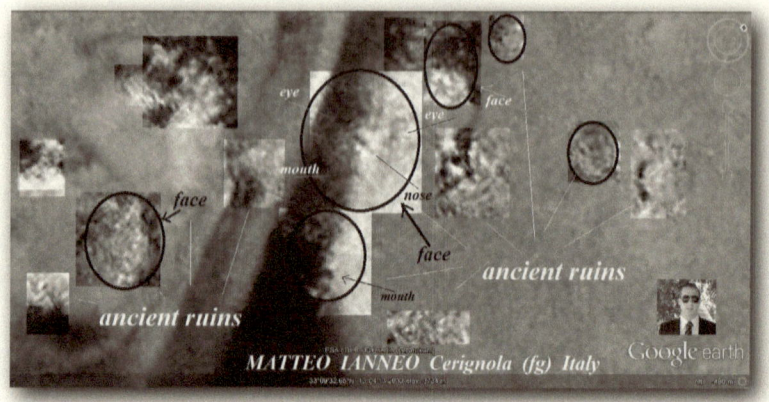

Image source: ESA/DLR/FU Berlin (G.Neukum)

Image source: ESA/DLR/FU Berlin (G.Neukum)

Other details

Image source: ESA/DLR/FU Berlin (G.Neukum)

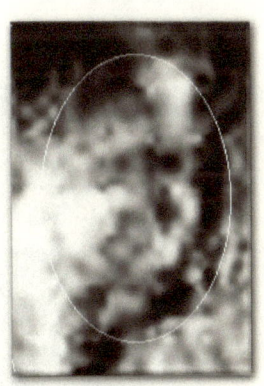

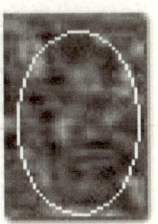

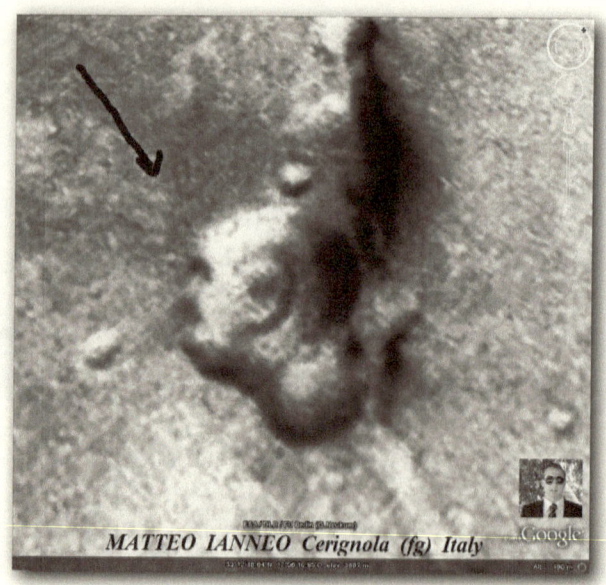

Image source: ESA/DLR/FU Berlin (G.Neukum)

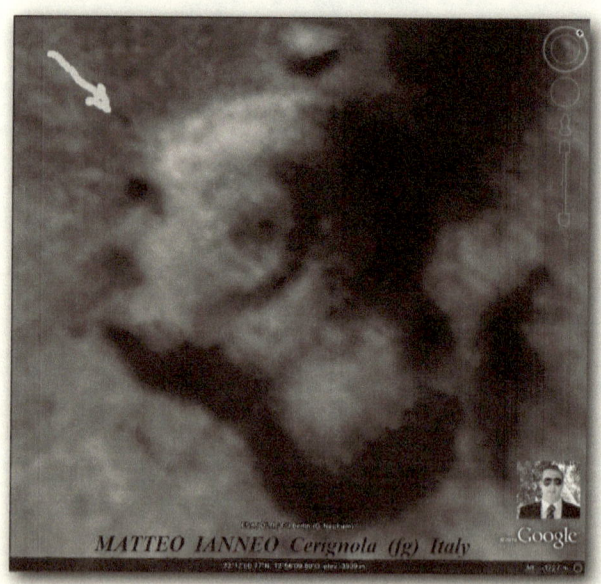

Image source: ESA/DLR/FU Berlin (G.Neukum)

With filter.

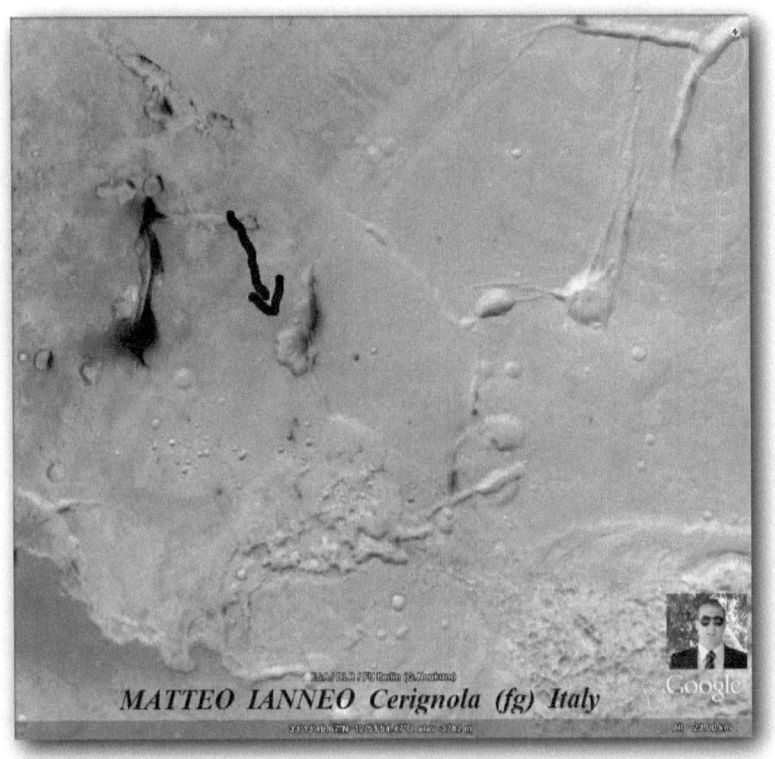

Image source: ESA/DLR/FU Berlin (G.Neukum)

This discovery I filed in a folder called *Secrets*. My curiosity grew hand in hand with my desire to explore. I never imagined that in just a few days I would add a second face. In the second face, also discovered in September 2009, the month of my wedding anniversary, it is possible to see the nose, and eyes with eyebrows; the mouth is not present, but is covered by something that I can't describe. It is a very beautiful face and seems like a painting on a canvas. When I reported the news to magazines, experts told me that this face was created by nature, and to our eyes it looks like something we have already seen somewhere on our own planet. I am not convinced. However, look at this picture and draw your own conclusions. I am sure others will have the same idea as me.

The position on
Google Earth is as follows:
Latitude 43°19'19.30"N   Longitude 22°53'5.29"E

Image source: ESA/DLR/FU Berlin (G.Neukum)

A beautiful face which in my opinion is feminine in nature.

If this face is artificial in nature, it must surely have been carved by some unknown means. My hypothesis is a sculpture carved from above by technology unknown to us. Perhaps enormous spaceships left these marks as a sign of their arrival on this planet – a type of competition to see who was the most powerful god explorer and conqueror of universes.

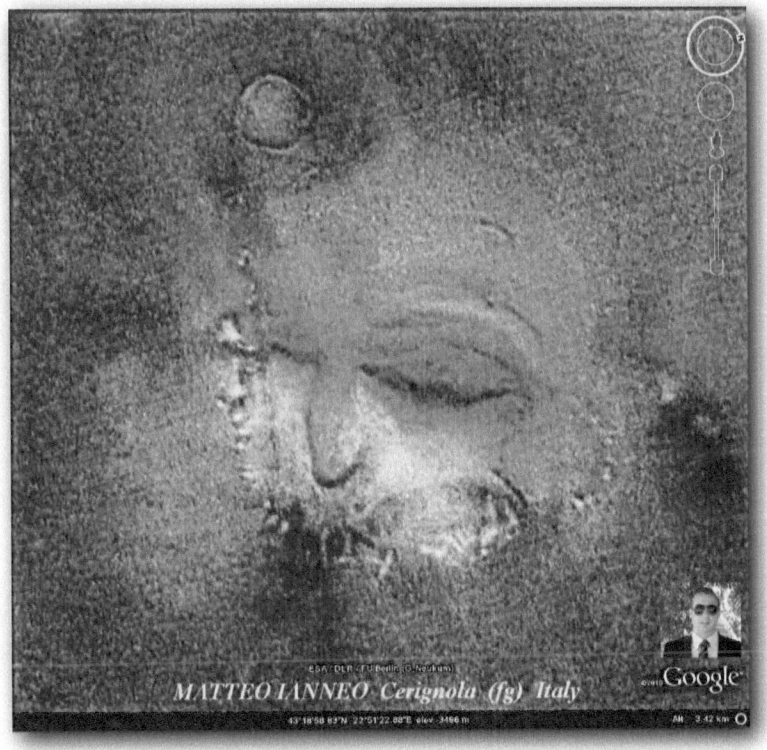

Image source: ESA/DLR/FU Berlin (G.Neukum)

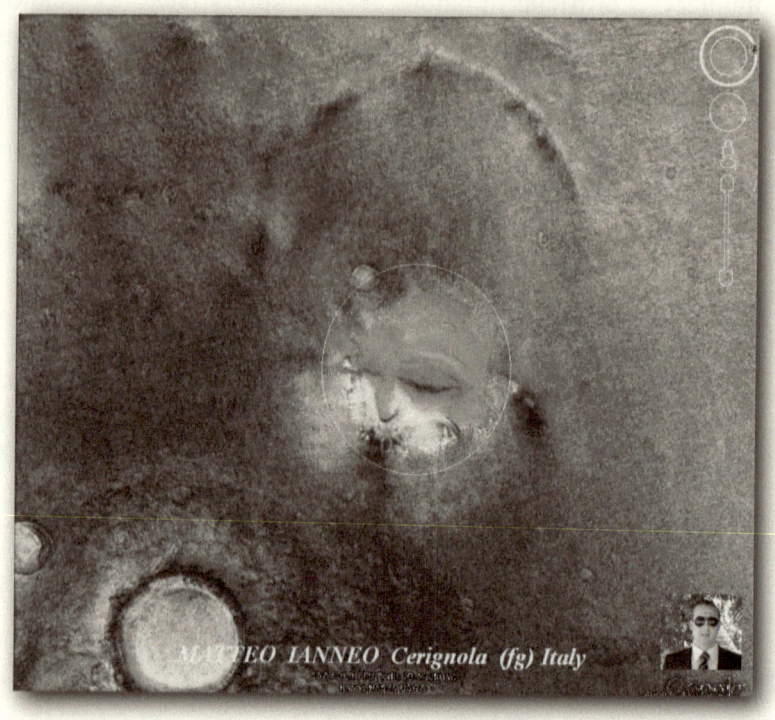

Image source: ESA/DLR/FU Berlin (G.Neukum)

Here we can see it from a different height. It can already be seen from this altitude that it is a face. Above the head a hidden circular shape can be seen, similar to a turban.

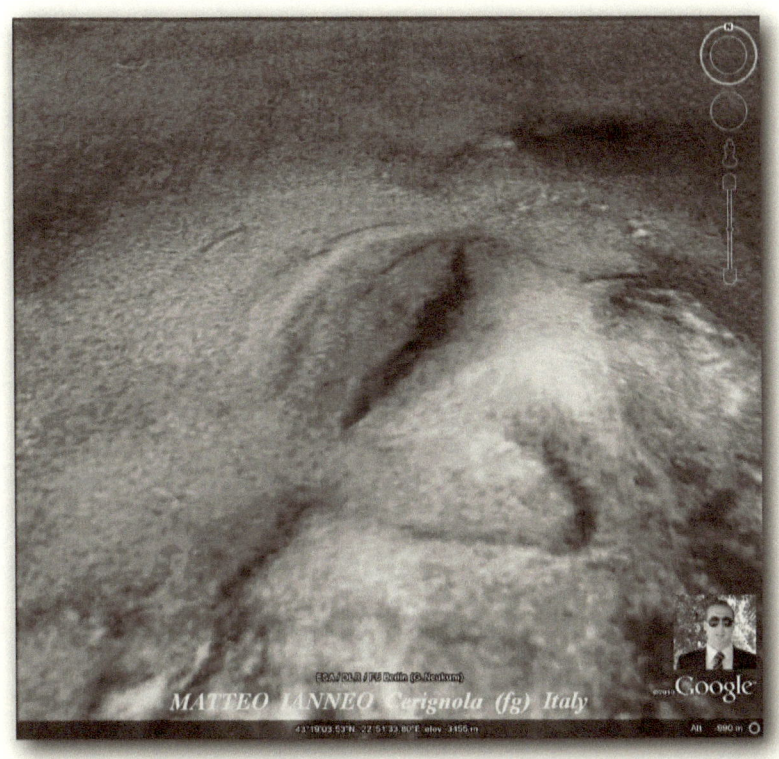

Image source: ESA/DLR/FU Berlin (G.Neukum)

Perspective 1

Image source: ESA/DLR/FU Berlin (G.Neukum)

Perspective 2

Image source: ESA/DLR/FU Berlin (G.Neukum)

Perspective 3

Image source: ESA/DLR/FU Berlin (G.Neukum)

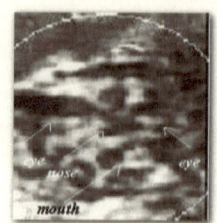

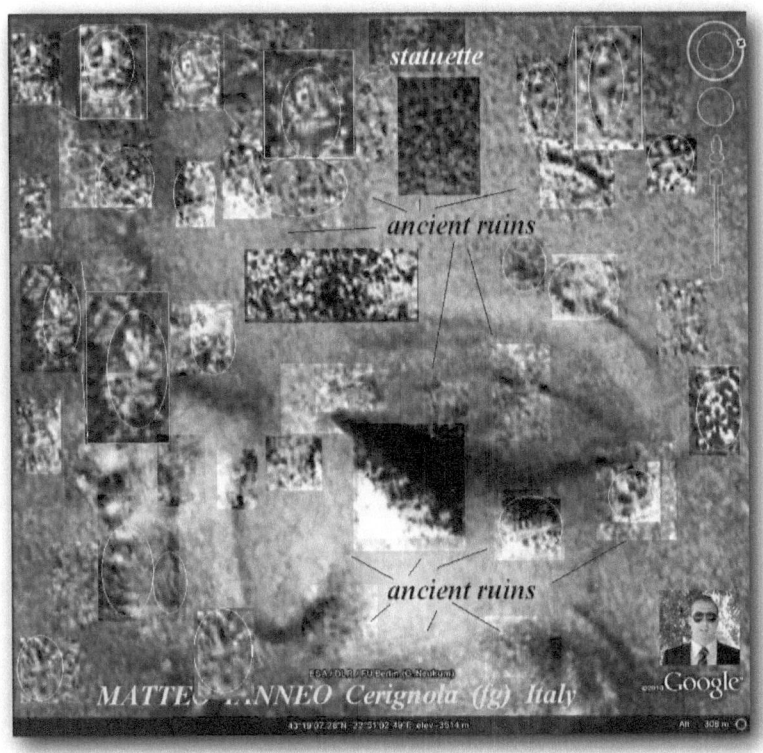

Image source: ESA/DLR/FU Berlin (G.Neukum)

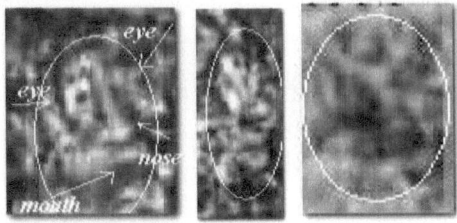

I kept thinking that what was amazing about these images was that they shared common details to us, i.e. human features. I wasn't surprised that later I discovered others. And so it was. As I scanned the surface of Mars, in one area I noticed the profile of an anthropomorphic being.

The position on

Google Earth is as follows:

Latitude 40°01'25.89"N   Longitude 8°58'21.53"W

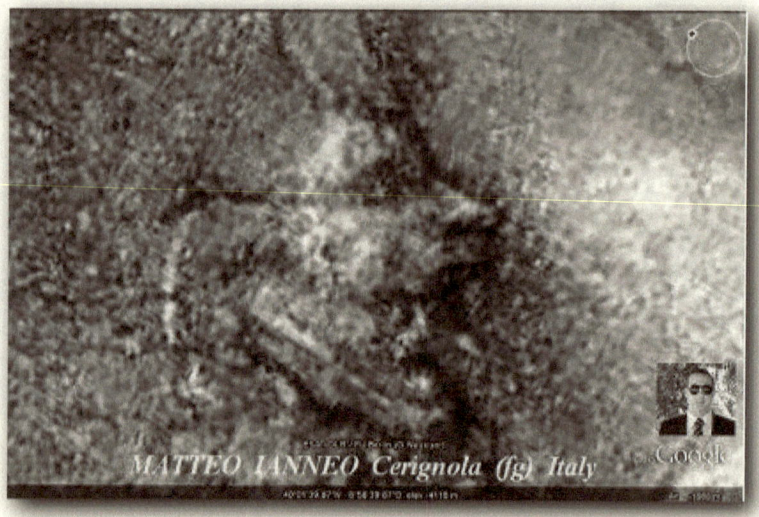

Image source: ESA/DLR/FU Berlin (G.Neukum)

It had the face of a wolf-man with its mouth open, as if it was shouting a war cry.

Image source: ESA/DLR/FU Berlin (G.Neukum)

If we look carefully, there is a snake coming out of his mouth, with eyes and mouth wide open. At the top right we can see the ruins of an ancient city. In the next photo we can see some features in more detail.

Image source: ESA/DLR/FU Berlin (G.Neukum)

At the top it is possible to distinguish a sphynx which has deteriorated over time. Note the head, the leg to the front, and the entrance to a cave, marked as entrance. Evidently life, at this time took place in caves, therefore, under the surface of Mars.

In the next sequence of photos we can see other details.

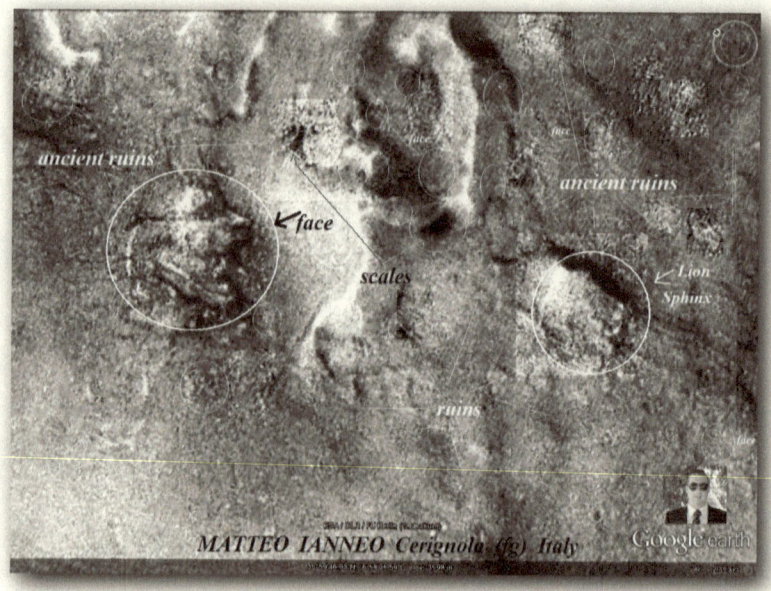

Image source: ESA/DLR/FU Berlin (G.Neukum)

I'll comment on the most evident details. To the right you can see the profile of a lion (sphynx), another face and steps which enter the cave. Without doubt many cities and many temples built on Mars were built with the aid of natural structures related to the formation of huge caves.

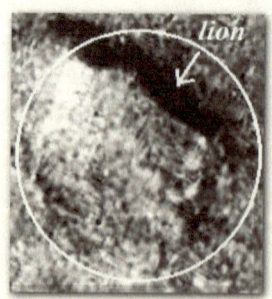

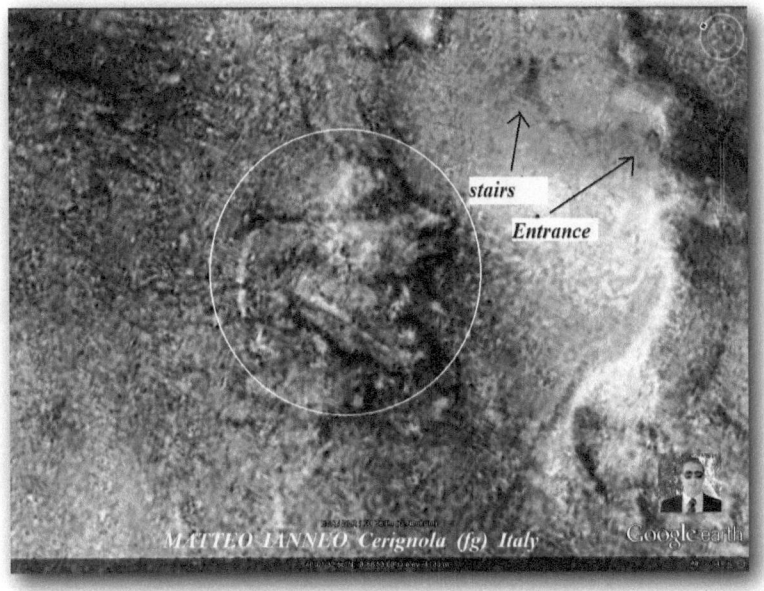

Image source: ESA/DLR/FU Berlin (G.Neukum)

In this image we can see the face of the main subject Anubis, the wolf god, surrounded to the top right by the city walls, with many entrances and steps for access. Without a doubt these ruins date back many many years. Mars, according to my hypothesis, was populated in the past by many different civilisations.

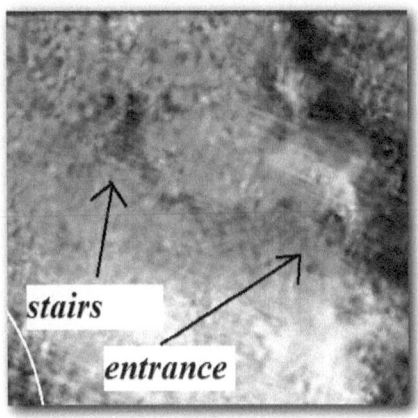

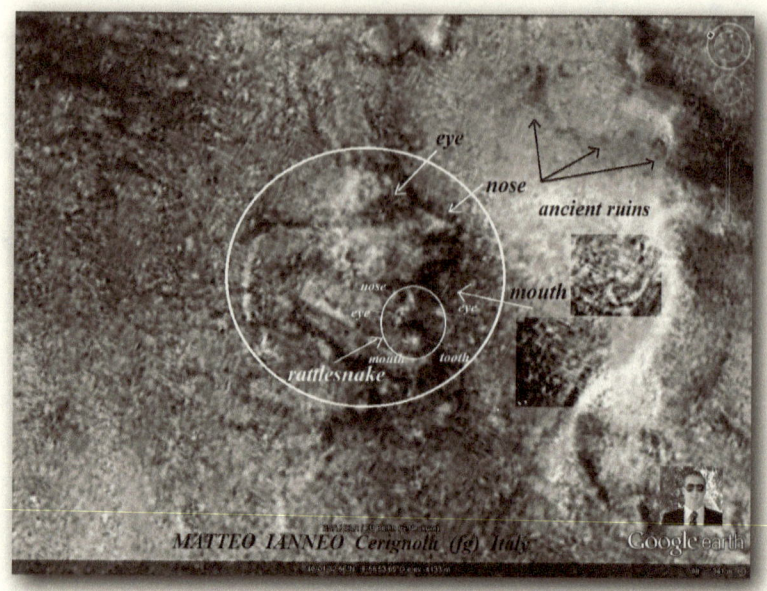

Image source: ESA/DLR/FU Berlin (G.Neukum)

Here we can see the reptile in the mouth of the main subject, with his mouth open. Looking at the mouth of the god Anubis, we note that very sharp teeth can also be seen. This may mean that this god was greatly feared for his bite, as strong as a wolf's, but poisonous as a rattle snake. Of course these are only my hypotheses.

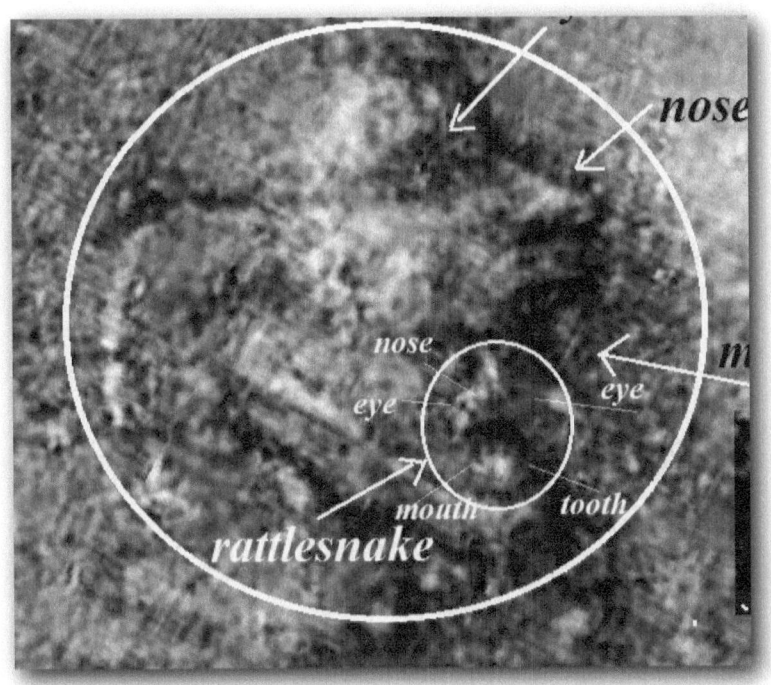

Image source: ESA/DLR/FU Berlin (G.Neukum)

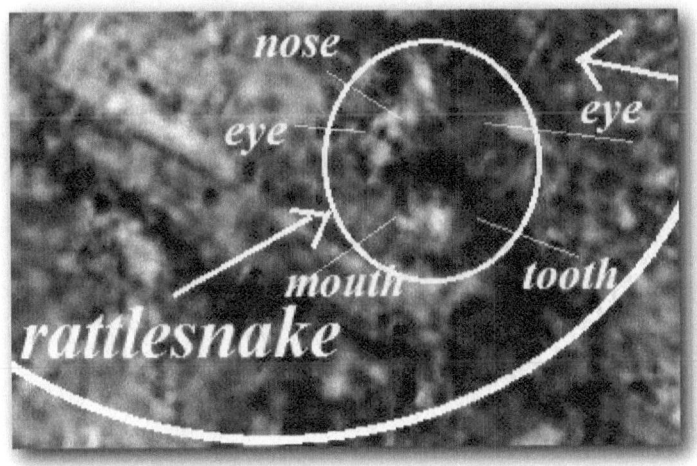

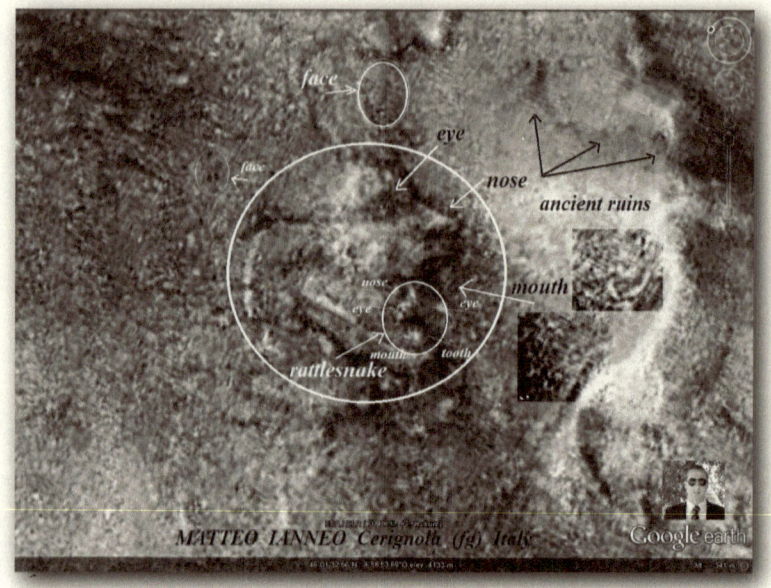

Image source: ESA/DLR/FU Berlin (G.Neukum)

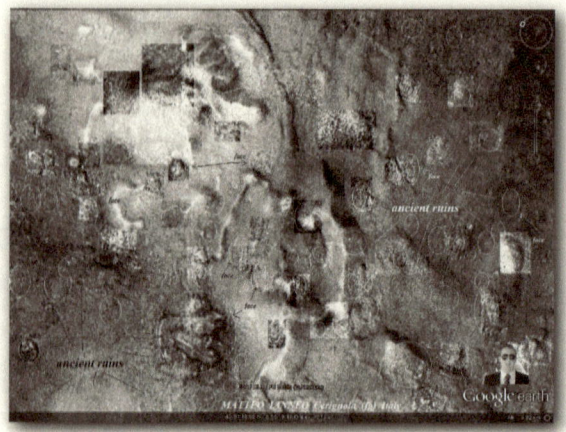

Image source: ESA/DLR/FU Berlin (G.Neukum)

On the right, in the middle of the photo, there is another face protruding from the rock, a little celebratory monument to someone from that history.

The position on
Google Earth is as follows:
Latitude 40°23'23.93"N   Longitude 9°33'12.80"W

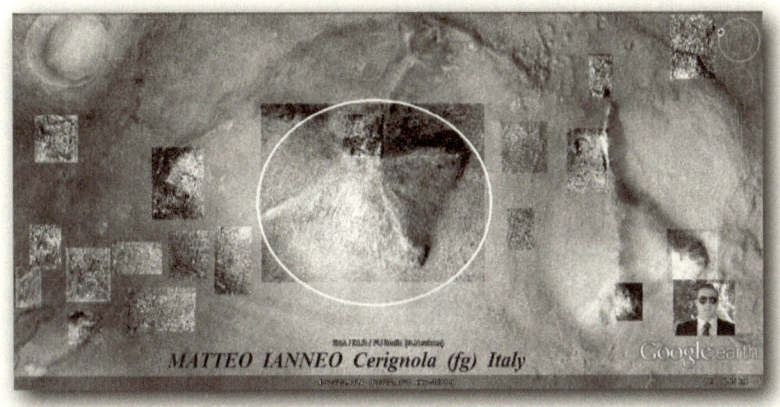

Image source: ESA/DLR/FU Berlin (G.Neukum)

A very ancient pyramid.

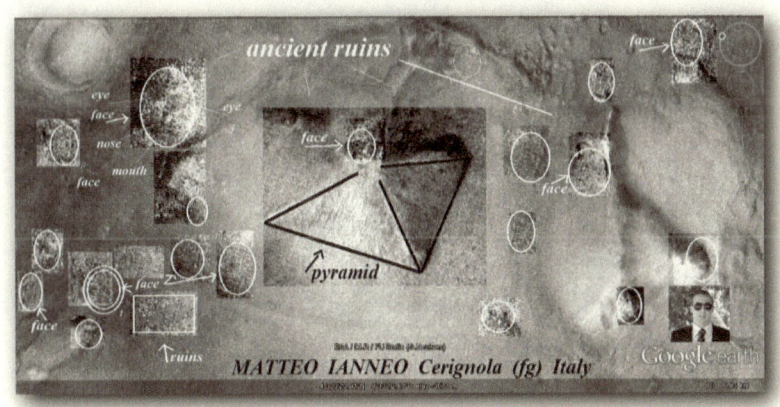

Image source: ESA/DLR/FU Berlin (G.Neukum)

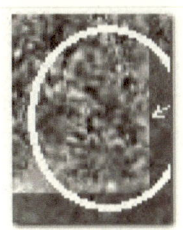

Based on these finding, I often wondered if one day I might find something more modern. A few days later I discovered some geometric structures.

Some hangars.

The position on
Google Earth is as follows:
Latitude 22°39'25.19"N   Longitude 103°47'56.61"W

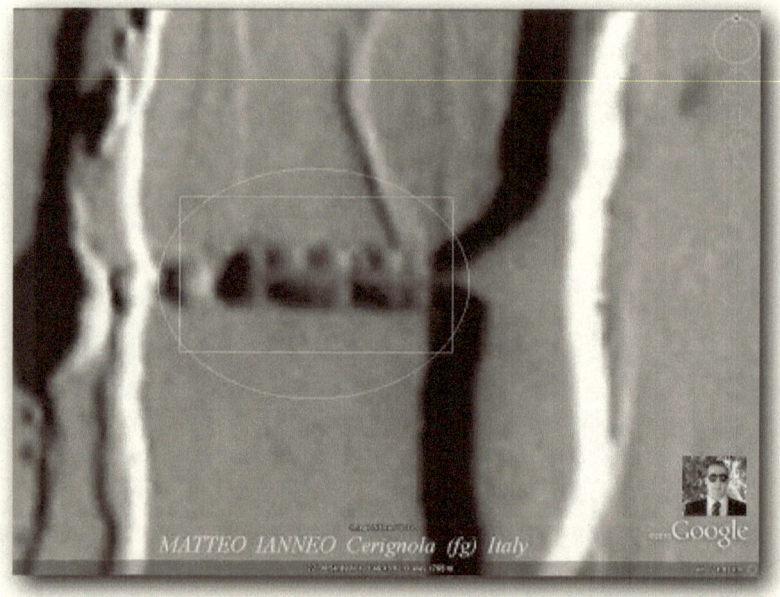

Image source: NASA/USGS

Some expert Mars scholars sent me emails in which they reiterated the theory that the details found by me are nothing more than natural rock formations as a result of water sources over time. They didn't convince me. If you look closely, the three geometric structures create parallels of 90° between them; it would be difficult for these to arise naturally.

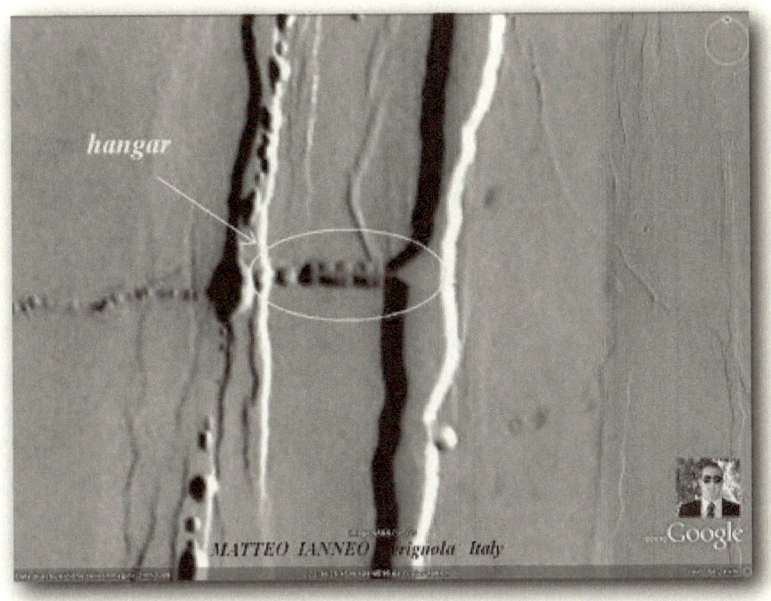

hangar

Image source: NASA/USGS

From this perspective we can see the geometric formations with parallel structures between them. These could be, according to my hypotheses, either access to underground (for example access for spaceships and spacecraft) or filtering holes used as ventilation ducts, i.e. intake and recycling of oxygen for the inhabitants living below the surface of the planet.

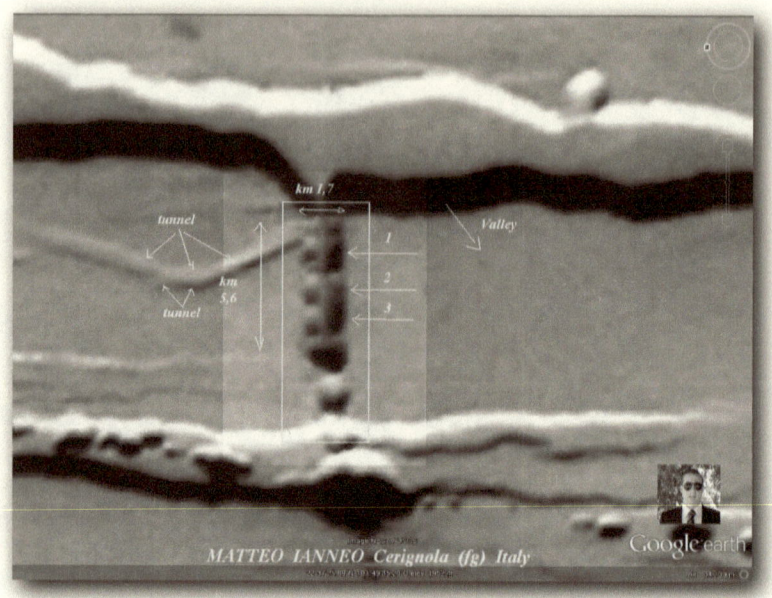

Image source: NASA/USGS

I inverted the structure to highlight the geometric parts. There are parallel rectangles between them, equidistant and the same shape. I do not believe that it is possible to find such things in nature, only constructions of artificial origin.

I thought about what else I might discover if I continued at this rate. I noticed, in another part of the planet, a fascinating example of a species of horse with legs, back, mane, and the face of a being different from that which I expected. Given that we are on a different planet, the forms depicted are likely to vary.

The position on

Google Earth is as follows:

Latitude 29°4'29.31"N   Longitude 60°8'56.87"W

Image source: NASA/USGS

Here is the photo of an animal with the face of a wolf, or of a bear. It is also possible to distinguish the hind legs and its mane.

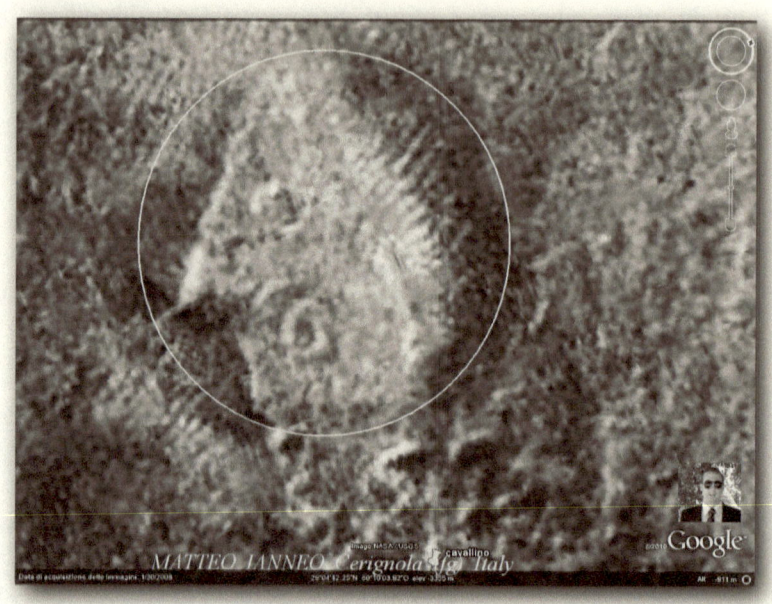

Image source: NASA/USGS

The head of the being highlighted.

Image source: NASA/USGS

Perspective 1

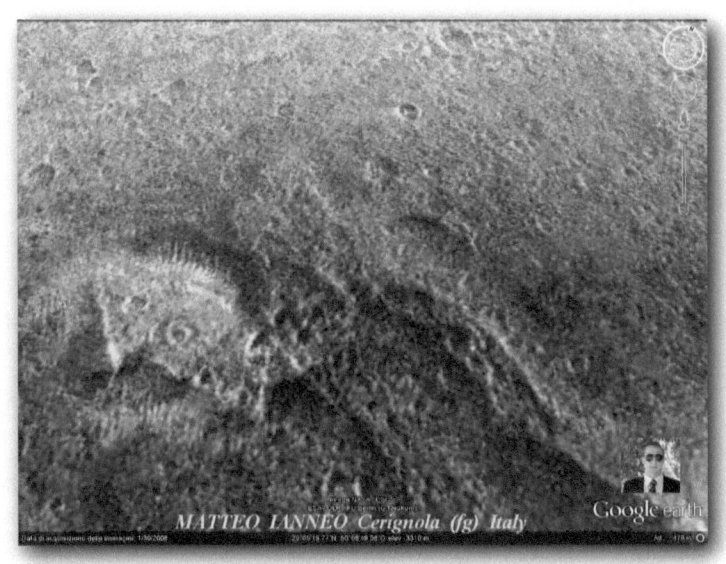

Image source: NASA/USGS

Perspective 2

Image source: NASA/USGS

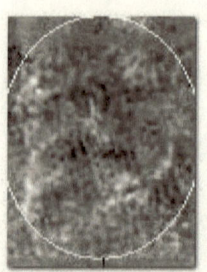

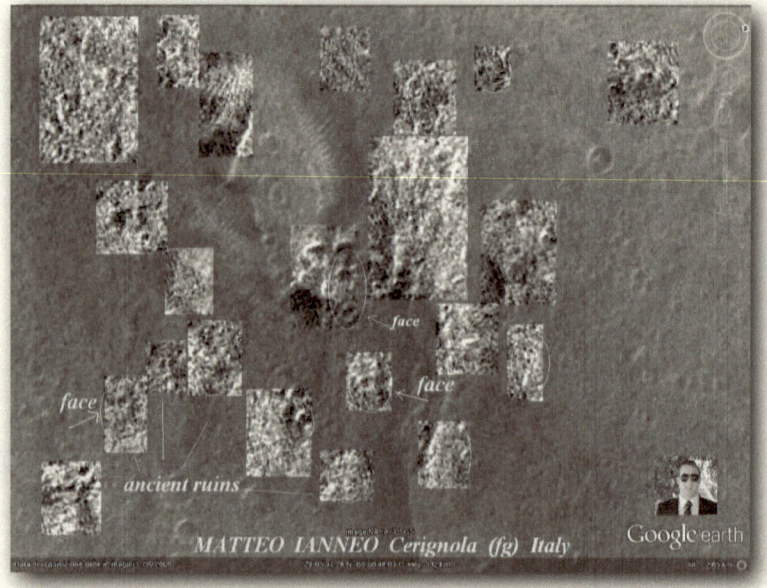

face

face

face

ancient ruins

MATTEO IANNEO Cerignola (fg) Italy

Google earth

Image source: NASA/USGS

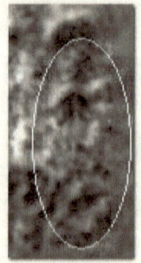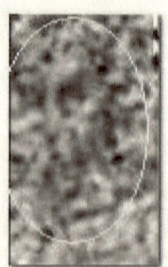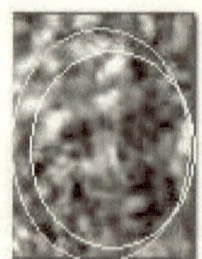

Then I found a representation of a monument buried over time. It is a statue with one arm raised like the Statue of Liberty in the United States.

The position on

Google Earth is as follows:

Latitude 19°32'45.13"N   Longitude 99°44'46.10"W

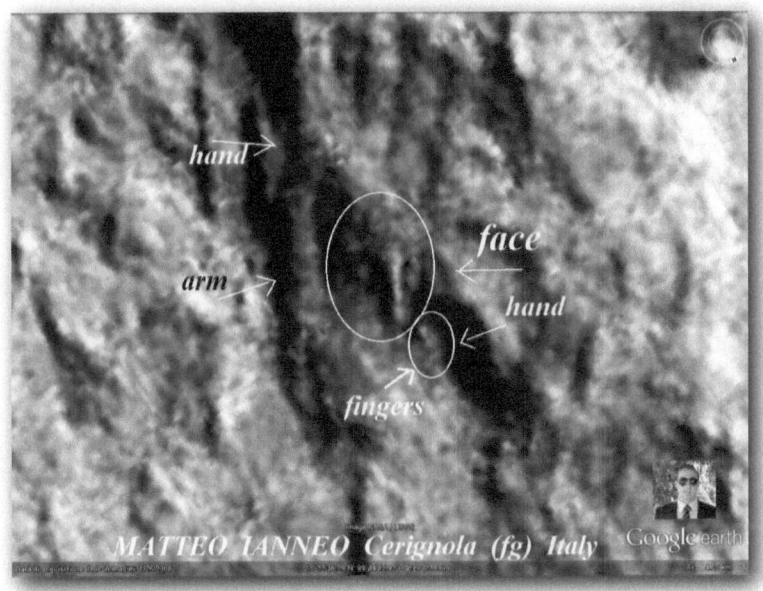

Image source: NASA/USGS

Here it is! A long nose, similar to Greek figures, one hand to the left side, you can see the fingers and the right arm raised as if he were holding something that no longer exists. Nearby it is possible to see other details.

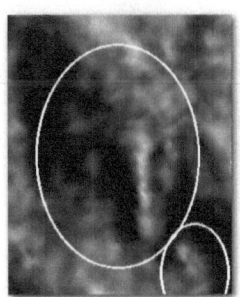

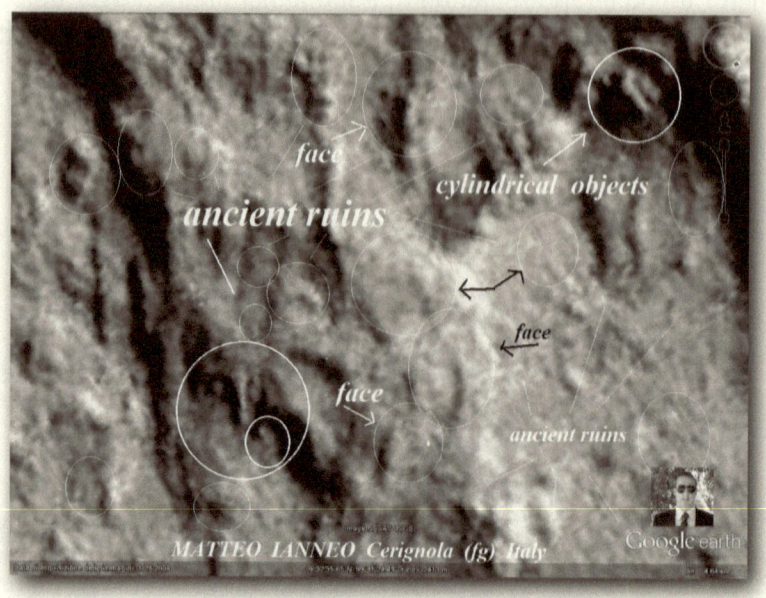

Image source: NASA/USGS

In the upper right there are some cylindrical objects. According to my hypotheses these could be silos abandoned for some time. Evidently someone was using these containers for fuel or food in order to stay for some time on the planet. Stacked cylinders, and then left there, provisions for some extraterrestrial aircraft during space missions carried out in the past.

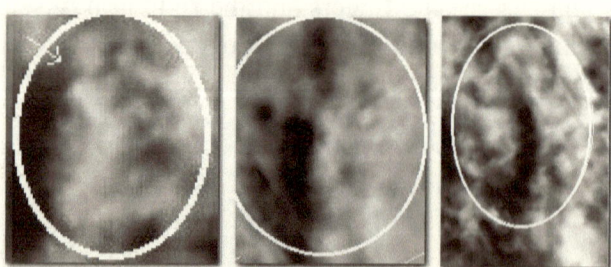

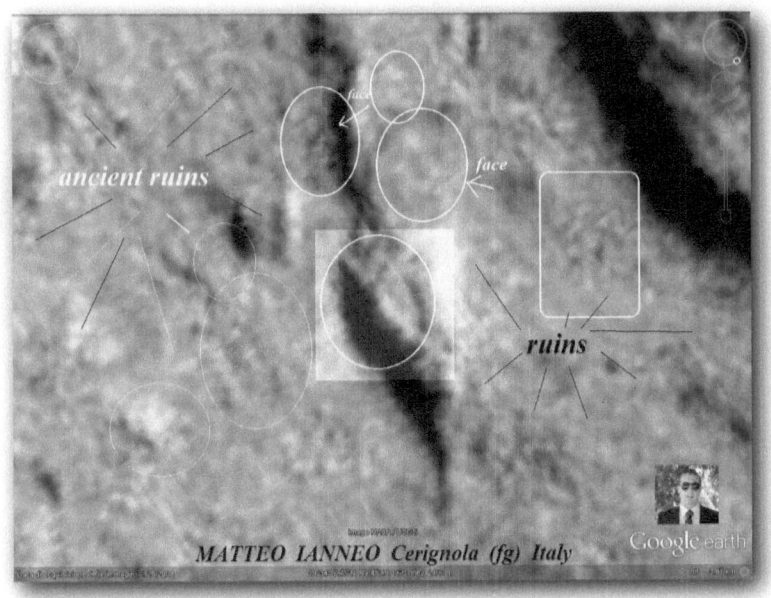

Image source: NASA/USGS

Another cylindrical formation.

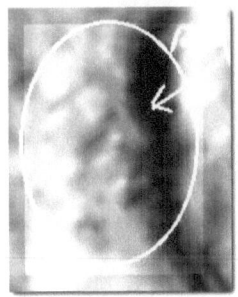

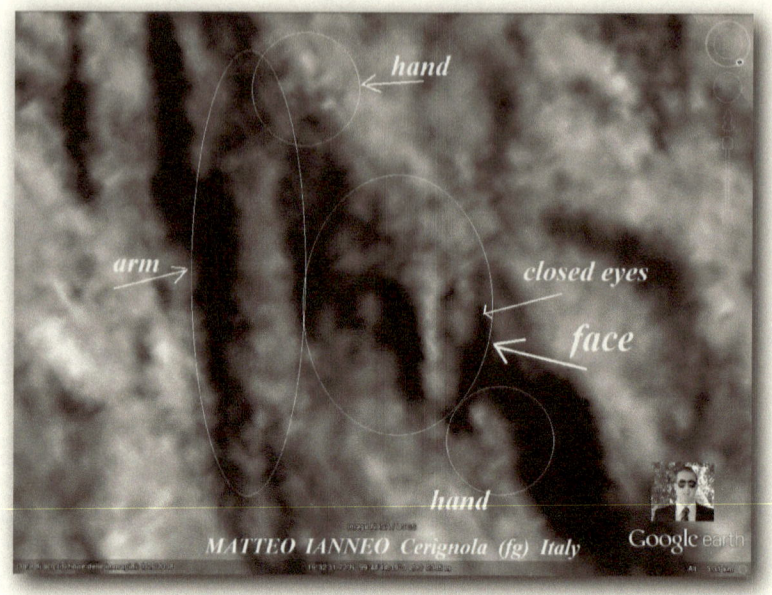

Image source: NASA/USGS

A face with closed eyes, a hand to the left side and an arm raised to the sky.

That same day I identified the face of an old man - a native face, a prehistoric man, the representation of something that caught my attention.

The position on

Google Earth is as follows:

Latitude 40° 6'4.28"N   Longitude 9°22'43.85"W

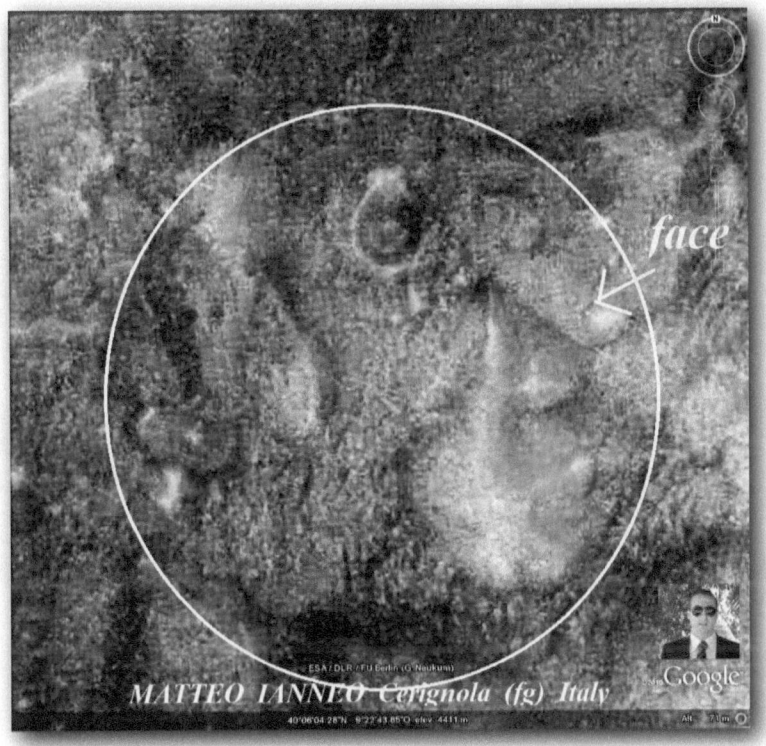

Image source: ESA/DLR/FU Berlin (G.Neukum)

You can see the mouth, nose, eye and teeth. Everyone asks "But are there only faces?" My reply is that around or inside those faces civilisation developed. The face was the biggest representation of their city. Their god.

Image source: ESA/DLR/FU Berlin (G.Neukum)

Here he is!

If I hadn't used my techniques it would have been difficult to find all this.

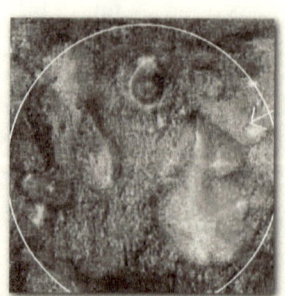

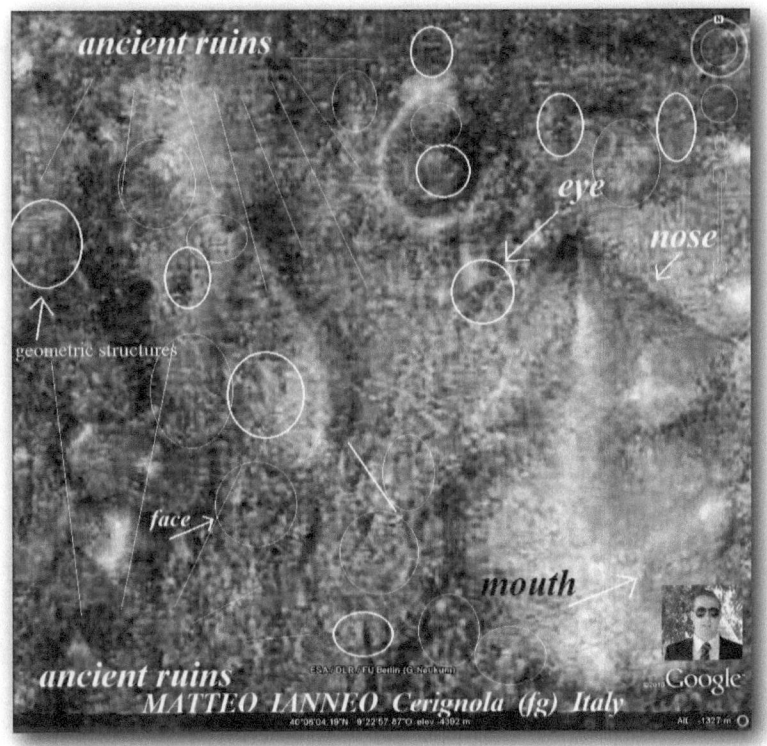

Image source: ESA/DLR/FU Berlin (G.Neukum)

The eye is circled and can be found beneath a small hut. You can see the iris. To the top left is the city. Geometric shapes which are accessed using numerous steps where the ancients carried their king on their shoulders. For devotion, replacing them with others, they transported their gods into the city hidden in the enormous caves. Here there would have been an altar in the centre where a priestess awaited the arrival of the king surrounded by the echoes of songs of a hundred voices accompanies by percussion instruments. The men, with powerful voices, produced sounds that resonated in the cave, combining the vibrations of their sounds with those of their spirits. "Of course, this is my imagination."

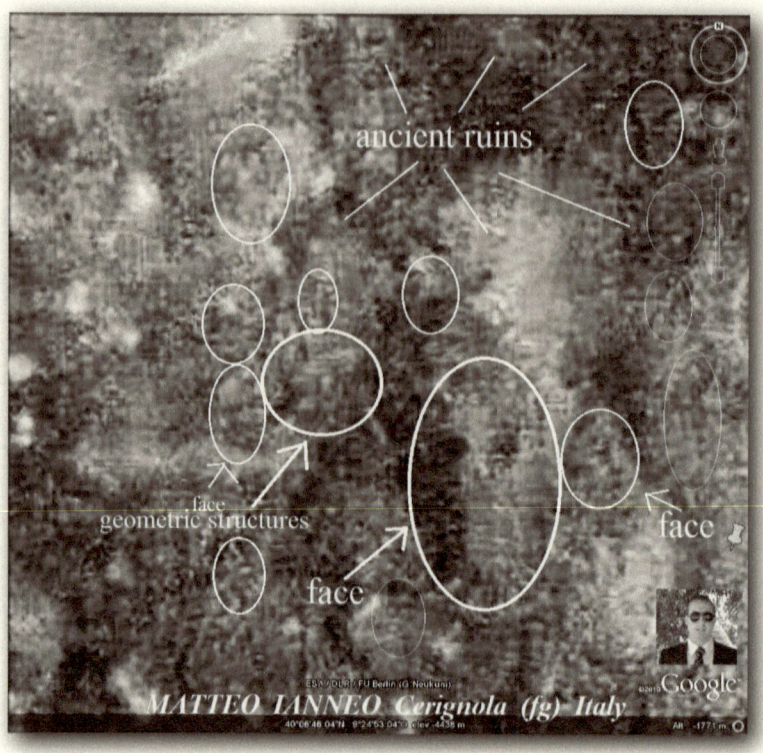

Image source: ESA/DLR/FU Berlin (G.Neukum)

Here is the city hidden in the caves, accessible using the long and numerous stairs. After having discovered them I asked myself: "Were there animals on Mars like those we find on Earth?"

The position on
Google Earth is as follows:
Latitude 50°15'2.86"N   Longitude 83°41'43.47"W

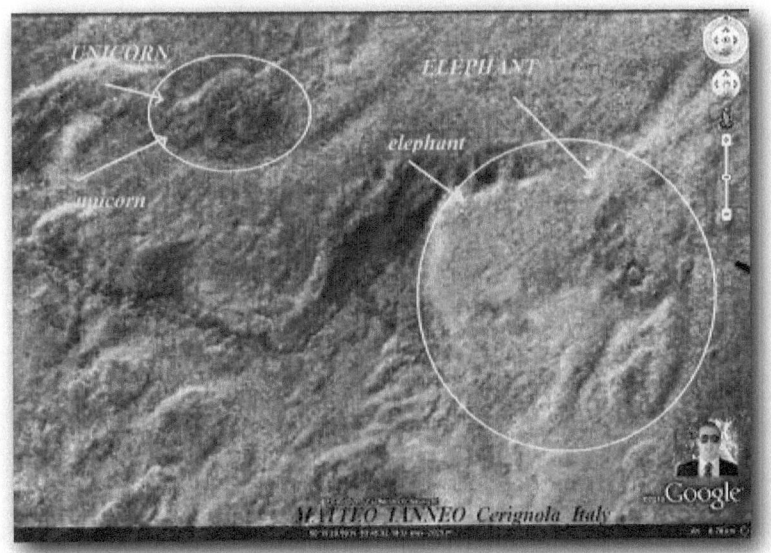

Image source: ESA/DLR/FU Berlin (G.Neukum)

Here I found an example of an elephant.

You can see an elephant with eyes, trunk, leg and back. To the top left we can see another animal, with a horn on its head: a goat, or perhaps a hare.

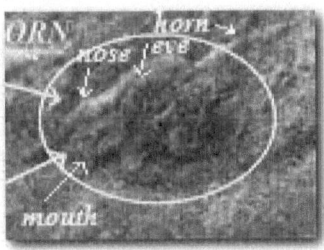

In ancient mythology animals often had horns on their heads. I found this also on Mars.

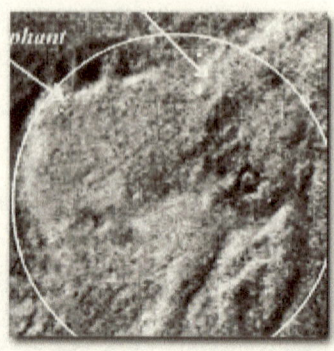

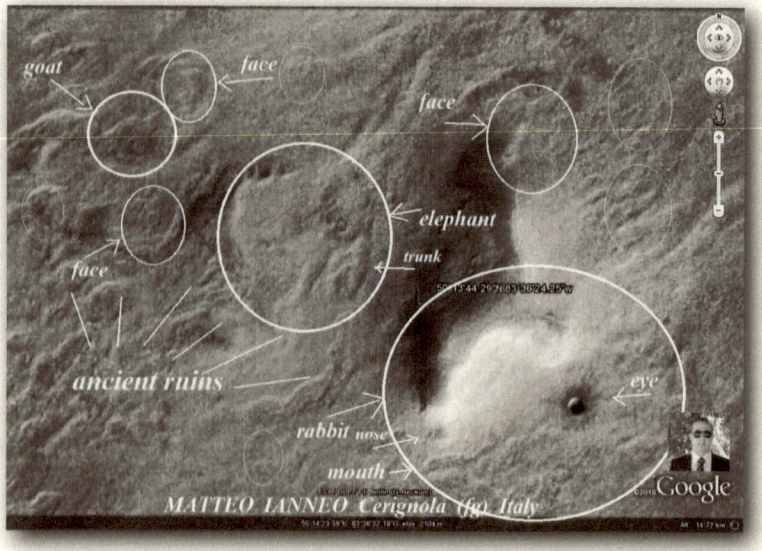

Image source: ESA/DLR/FU Berlin (G.Neukum)

Below right we see the head of a rabbit, and above it the profile of a face similar to that of a monkey. So an elephant, a rabbit and a goat with a horn on its head and the profile of a monkey on the mountain.

In ancient Egypt they worshiped the wolf-dog and I found this likeness.

The position on

Google Earth is as follows:

Latitude 61°51'01.62"S   Longitude 176°12'42.07"W

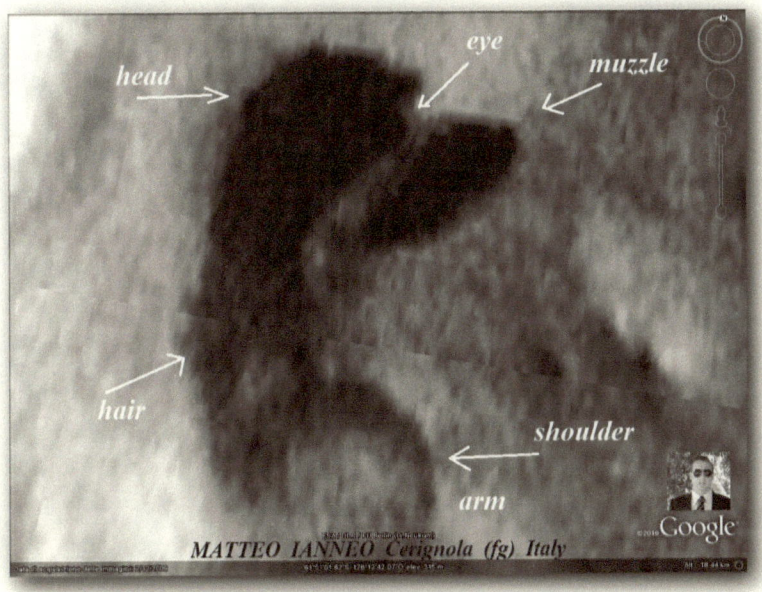

Image source: ESA/DLR/FU Berlin (G.Neukum)

It's possible to see the nose, an eye, his head, his hair which falls over his shoulders and right arm. This image contains within itself the history of a civilisation.

Image source: ESA/DLR/FU Berlin (G.Neukum)

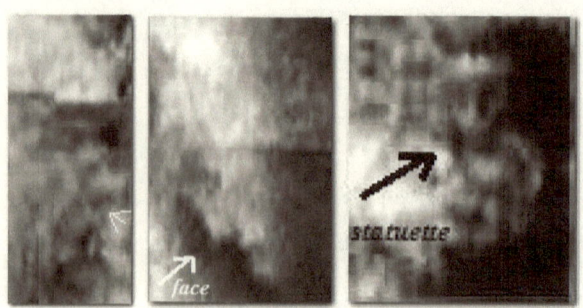

The position on

Google Earth is as follows:

Latitude 11°10'30.37"N   Longitude 104°26'40.81"W

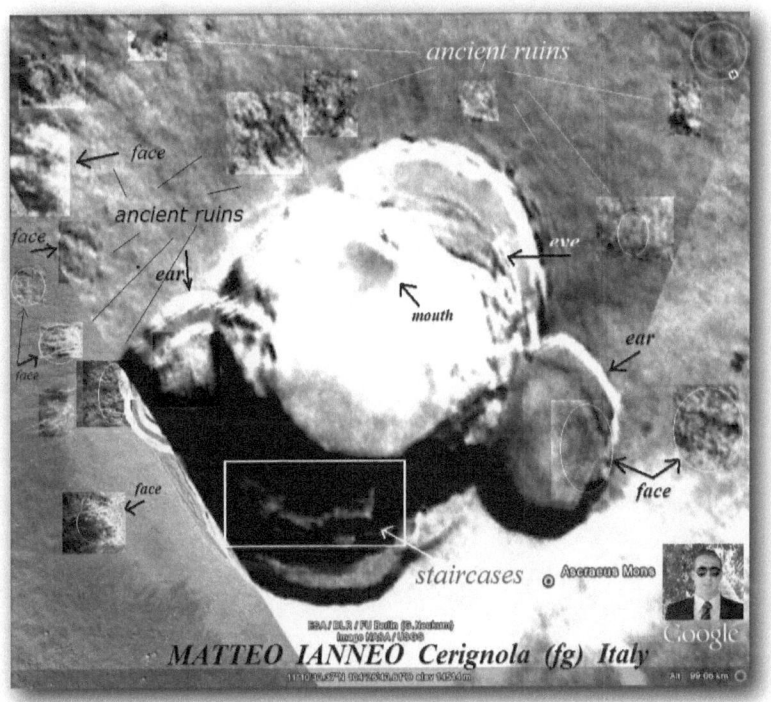

Image source: ESA/DLR/FU Berlin (G.Neukum)

NASA /USGS

In this frame we can see the access to a hidden city.

We can see a face looking up to the sky. It seems to have lips which have been immortalised in the act of giving a kiss. The ears and eyes seem to be oriental.

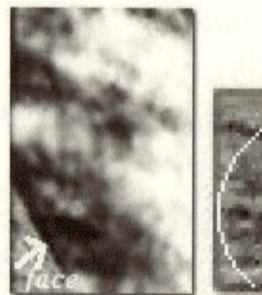

The vegetation is slowly taking shape. Someone is reviving what was once life on Mars. You can see in the images that there are some large staircases. According to my hypotheses, these are the remains of the walls of an ancient civilisation.

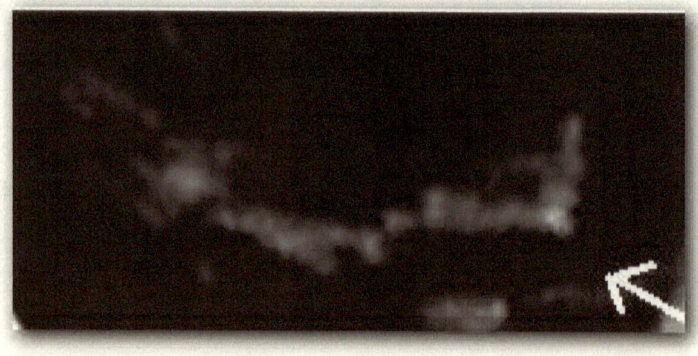

Detailed view.

(Ruins hidden below the surface)

Evidently the surface of Mars was not very hospitable in the past, perhaps because of some cataclysm, or the fear of invasion from above.

I wanted to observe the first face, I mean the one known throughout the world and situated in the region of Cydonia. I realised that it had been unmade, as if someone had manipulated the image, but I noticed something fascinating in its vicinity, something wonderful: the profile of a face enclosed in a pyramid. Extraordinary!

The position on

Google Earth is as follows:

Latitude 41°19'53.75"N Longitude 9°48'46.20"W

Image source: ESA/DLR/FU Berlin (G.Neukum)

Here we see a sphinx, a face depicted in a pyramid, where to the top right we can see – even if it is with little clarity – a type of cow or bull, while in front of the face of the sphynx we can see a snake, a cobra. In Egyptian worship these animals are often treated as gods, Ra Althor etc. Cobra, sphinx, cow or bull. Without doubt some experts on ancient history will know better than I what the link is between these elements.

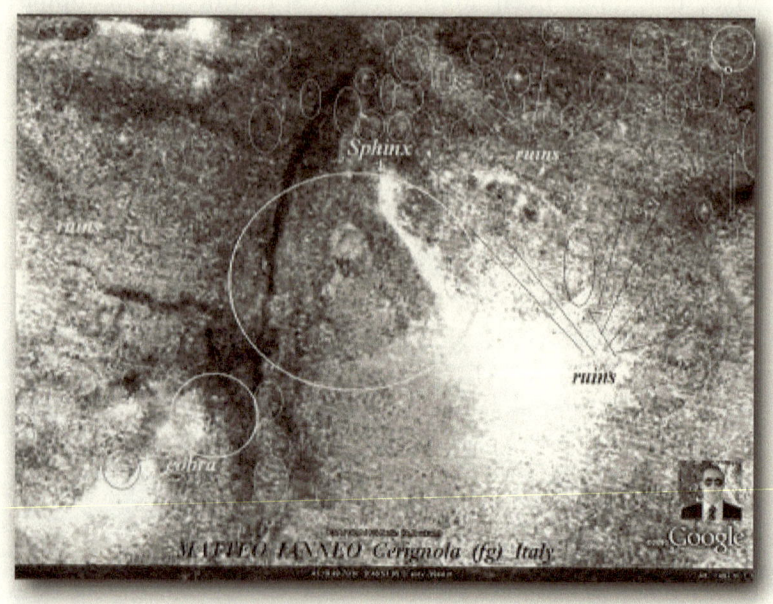

Image source: ESA/DLR/FU Berlin (G.Neukum)

The cobra to the bottom left.

The face at the top in the centre.

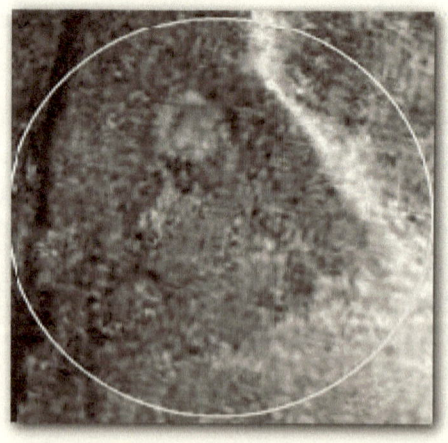

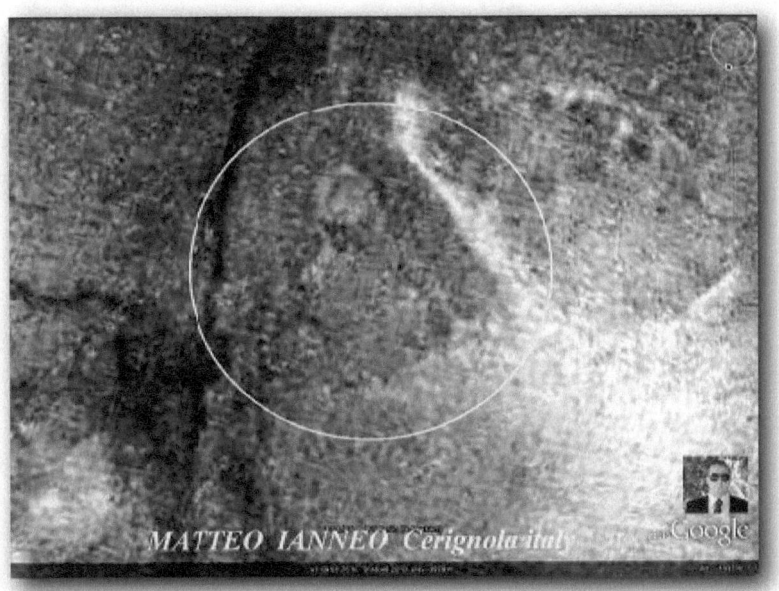

Image source: ESA/DLR/FU Berlin (G.Neukum)

If you can focus near to this face there are the ruins of a city, where the faces are not due to pareidolia, but are real monuments of these people, a real ancient city.

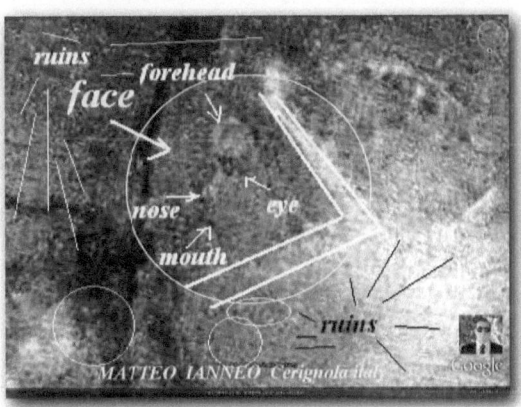

Image source: ESA/DLR/FU Berlin (G.Neukum)

You can see a man's face with a pronounced forehead, a pig's nose and a mouth with a long goatee beard. This is what my eyes see.

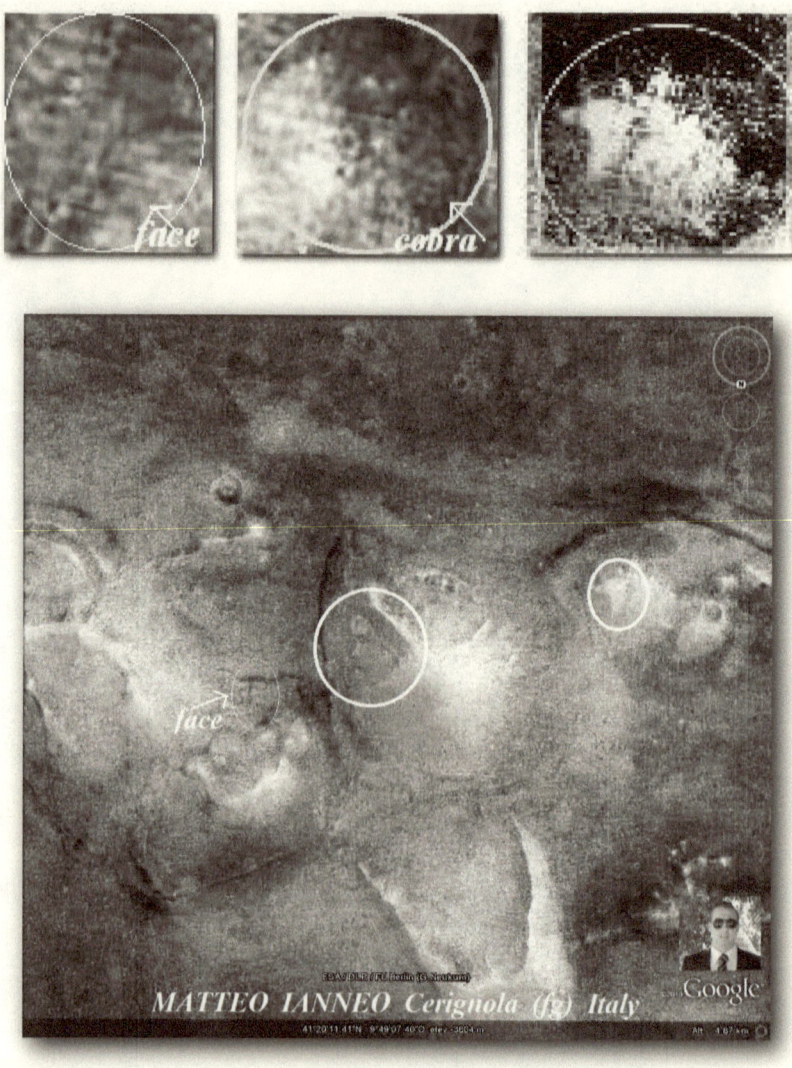

Image source: ESA/DLR/FU Berlin (G.Neukum)

Filtered.

Image source: ESA/DLR/FU Berlin (G.Neukum)

Original view of the image untreated by filters.

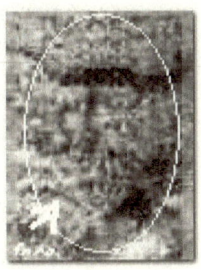

As you can see, if you don't have a good eye these details can be missed, also because they are not easy to find. It has taken me a long time to analyse these elements which, after many personal assessments, seem fundamental and representative of an ancient history unknown to us.

I have found many different faces, but this one really struck me for what it represents. A face of a particular being, placed in a triangular structure.

The position on

Google Earth is as follows:

Latitude 42°19'36.80"N   Longitude 27°53'54.38"W

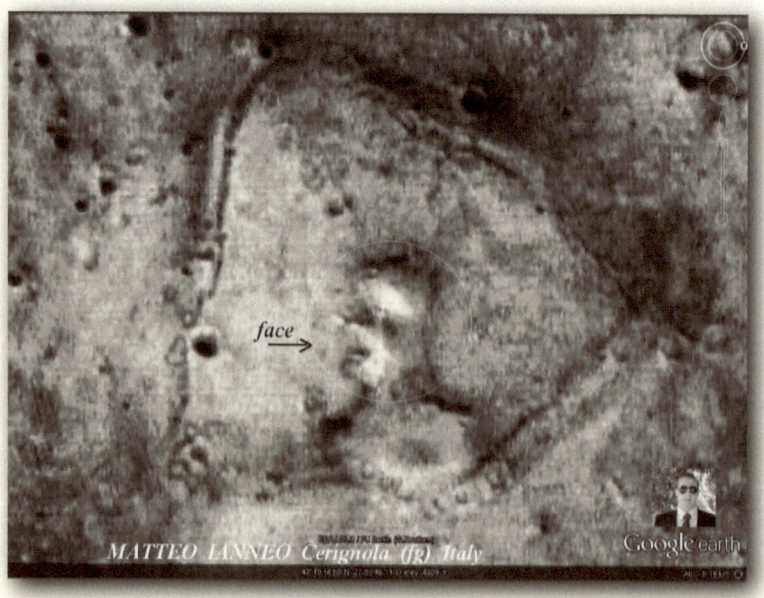

Image source: ESA/DLR/FU Berlin (G.Neukum)

It seems to be posed in order to give us a photo.

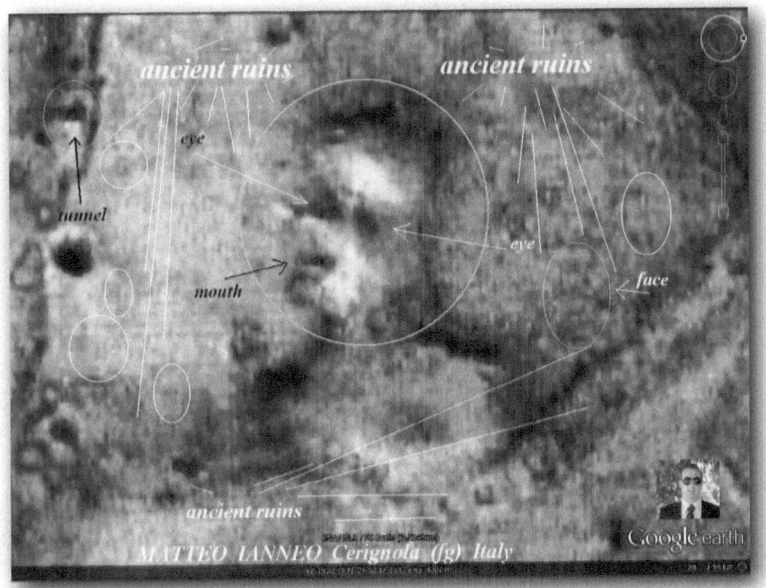

Image source: ESA/DLR/FU Berlin (G.Neukum)

A gaze which looks at us and an open mouth, as if to tell us something.

We can see a type of entrance to the left, a tunnel.

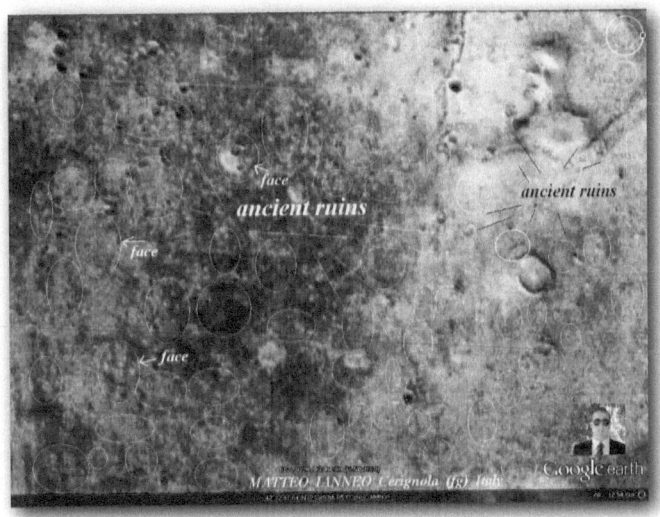

Image source: ESA/DLR/FU Berlin (G.Neukum)

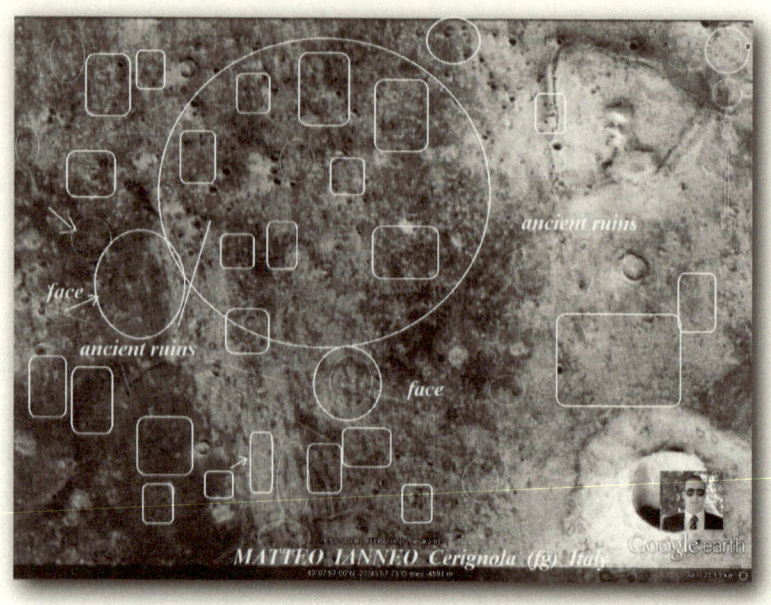

Image source: ESA/DLR/FU Berlin (G.Neukum)

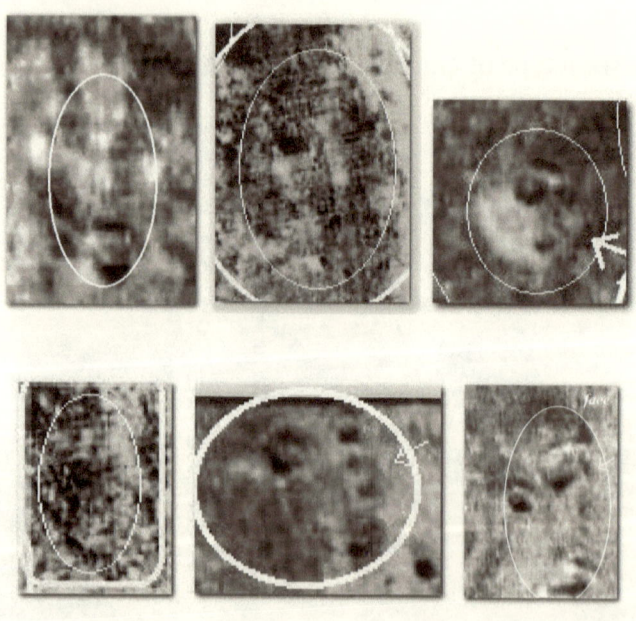

Image source: ESA/DLR/FU Berlin (G.Neukum)

This is a close-up. We can see the expression with open mouth, chin and lower lip.

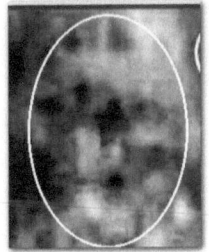

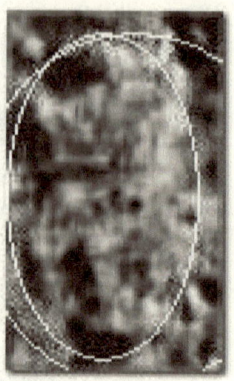

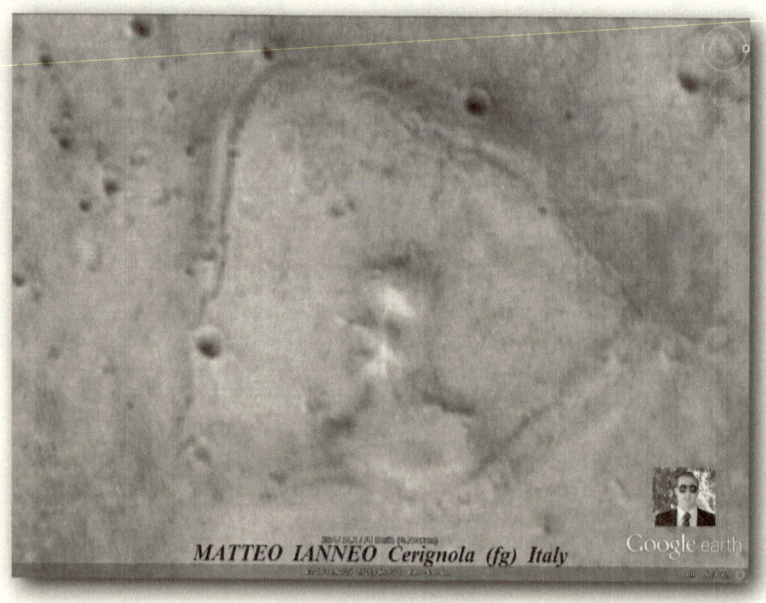

MATTEO IANNEO Cerignola (fg) Italy

Image source: ESA/DLR/FU Berlin (G.Neukum)

The original.

The position on
Google Earth is as follows:
Latitude 42° 9'10.39"N   Longitude 27°44'21.45"W

Image source: ESA/DLR/FU Berlin (G.Neukum)

Image source: ESA/DLR/FU Berlin (G.Neukum)

Ancient ruins.

Image source: ESA/DLR/FU Berlin (G.Neukum)

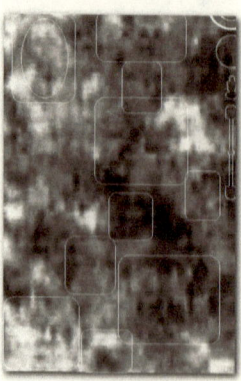

One detail I did not expect to find is this sculpture in the form of a tapir which I wanted to call *The Temple*.

The position on
Google Earth is as follows:
Latitude 40°19'54.40"N   Longitude 53°41'54.00"W

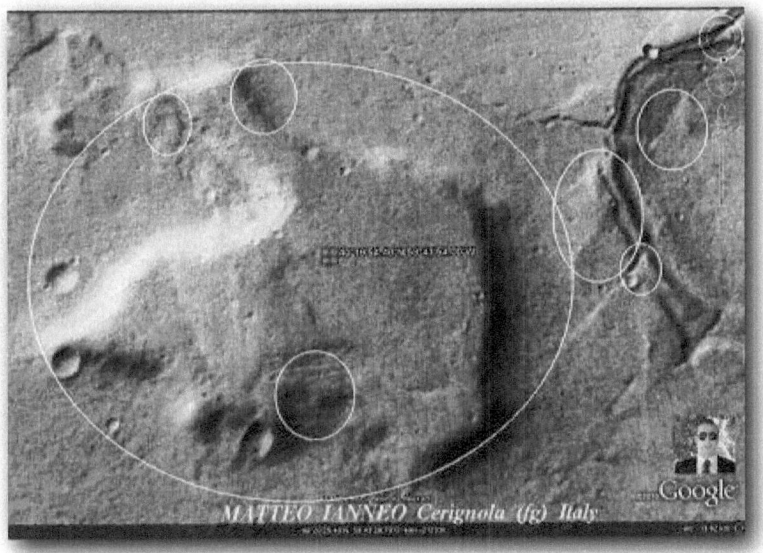

Image source: ESA/DLR/FU Berlin (G.Neukum)

In this image we can see some big stairs to the bottom right of
The Temple. You can also see profiles with a similarity, which
makes me think of the ancient Sumerians.

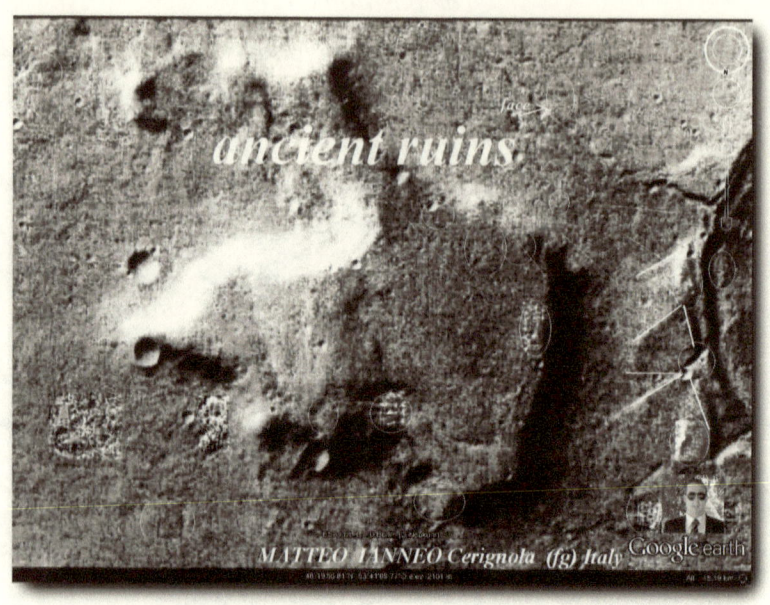

Image source: ESA/DLR/FU Berlin (G.Neukum)

To the top right we can find two pyramids next to the profiles.

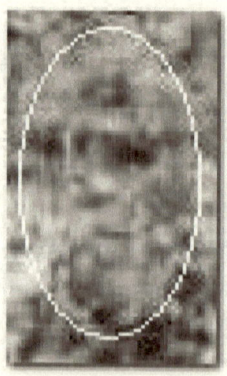

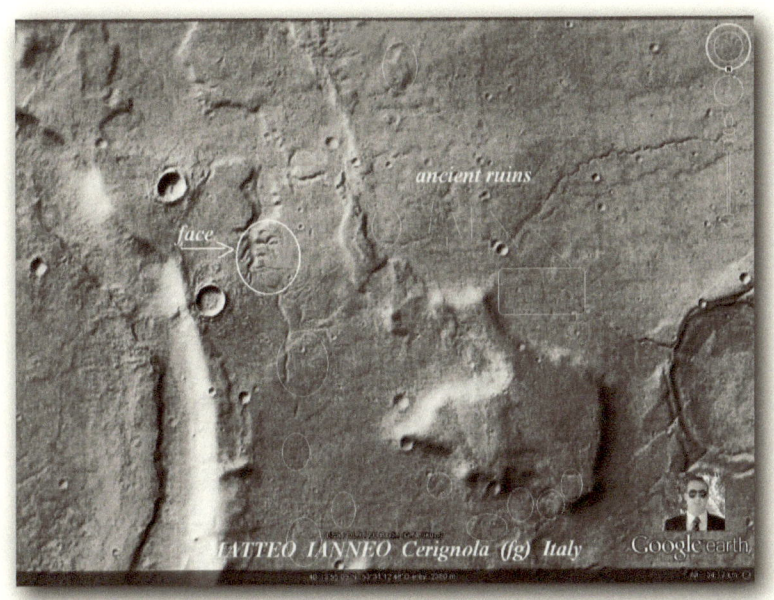

Image source: ESA/DLR/FU Berlin (G.Neukum)

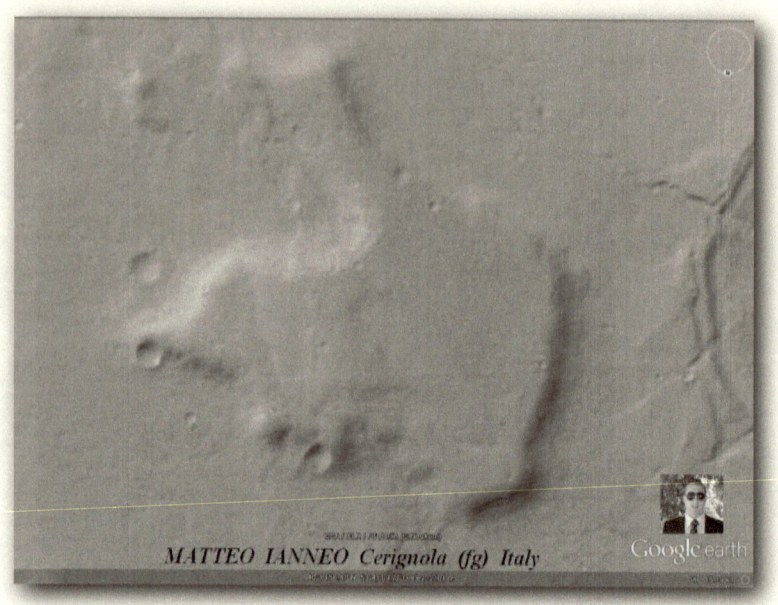

Image source: ESA/DLR/FU Berlin (G.Neukum)

The original.

As I continued to explore other places on the planet, I found another detail, a profile of a sculpture like the ancient Egyptian Sphynx.

The position on

Google Earth is as follows:

Latitude 10°13'3.38"S   Longitude 120°32'16.11"W

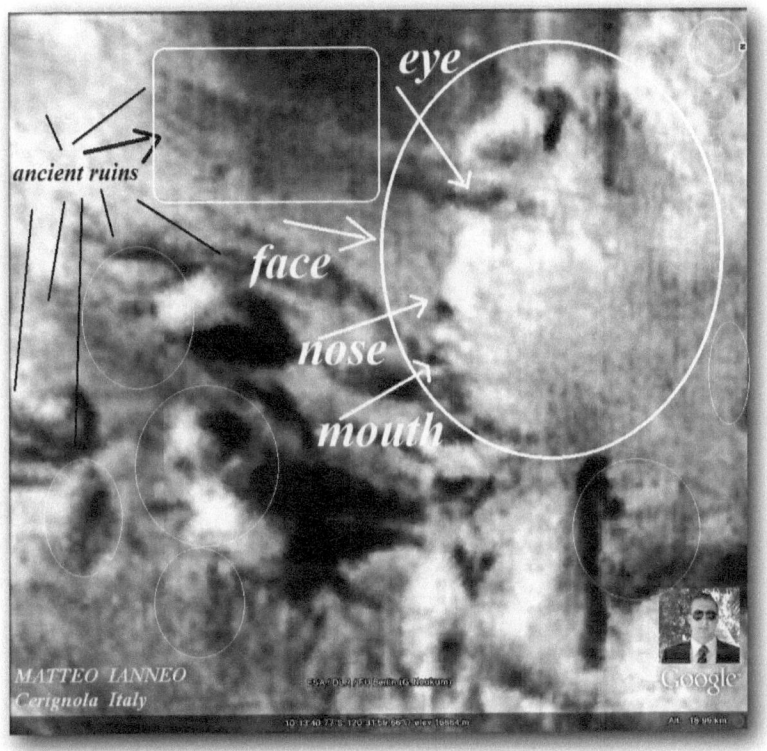

Image source: ESA/DLR/FU Berlin (G.Neukum)

Looking to the top right, we can see the face of a person with pronounced lips; he has the look of a powerful being with an angry face.

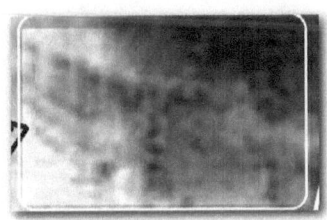

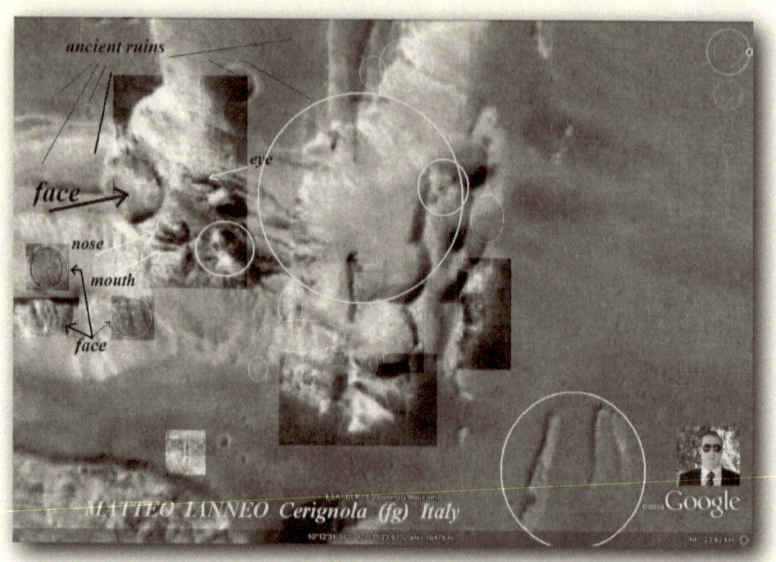

Image source: ESA/DLR/FU Berlin (G.Neukum)

Note, at the bottom, representations of designs made by people from the past. You can see a face carved in the rock, which probably represents the place where an ancient civilisation lived in this area of the planet.

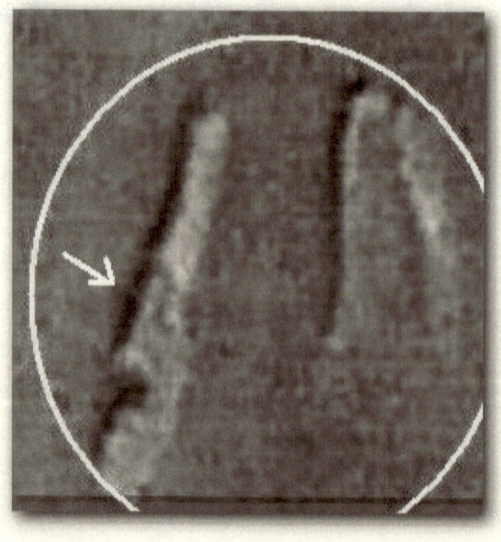

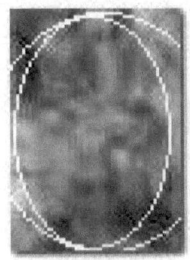 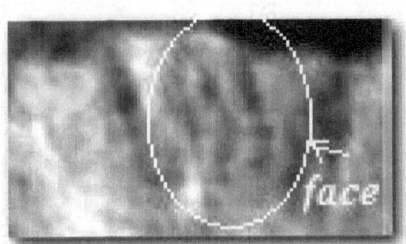

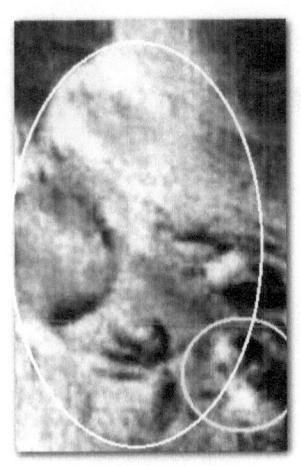

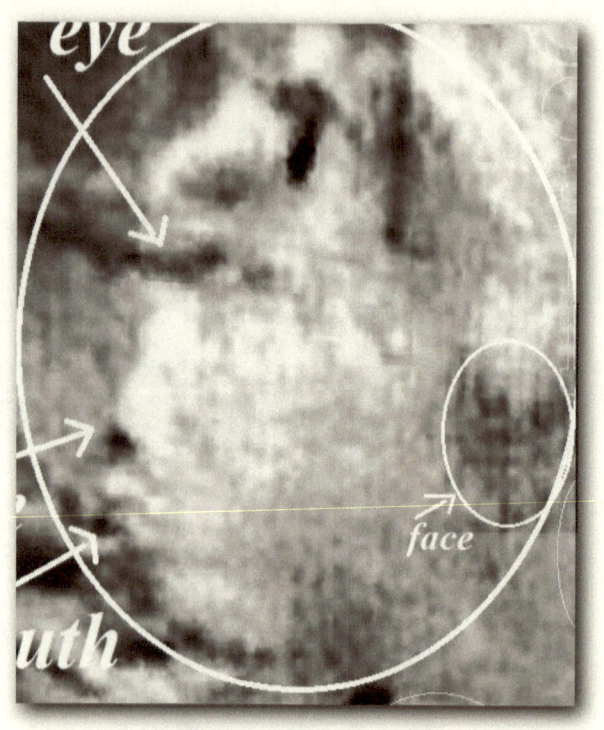

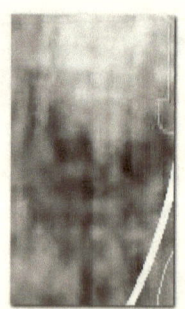

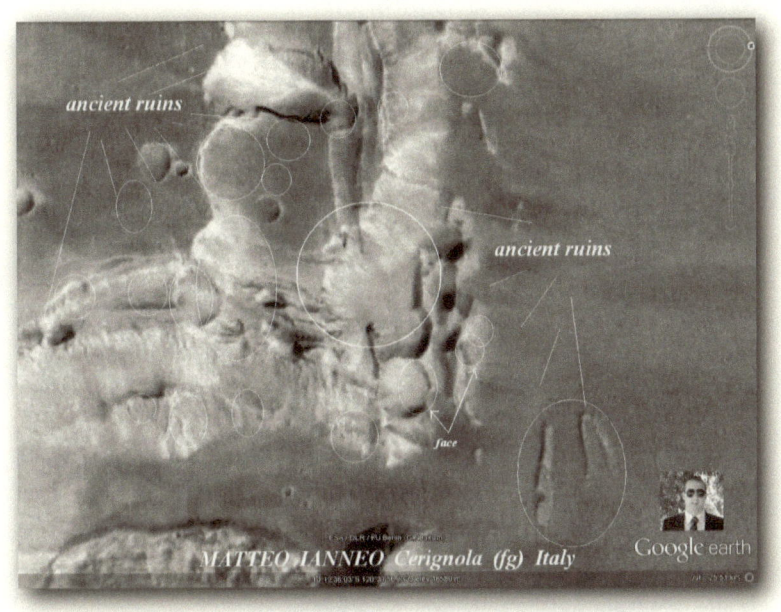

Image source: ESA/DLR/FU Berlin (G.Neukum)

The original.

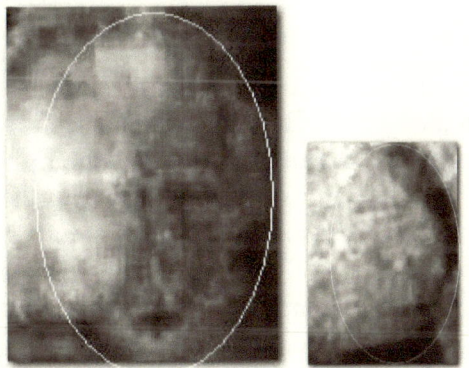

Then I discovered a sensational image, an important detail.

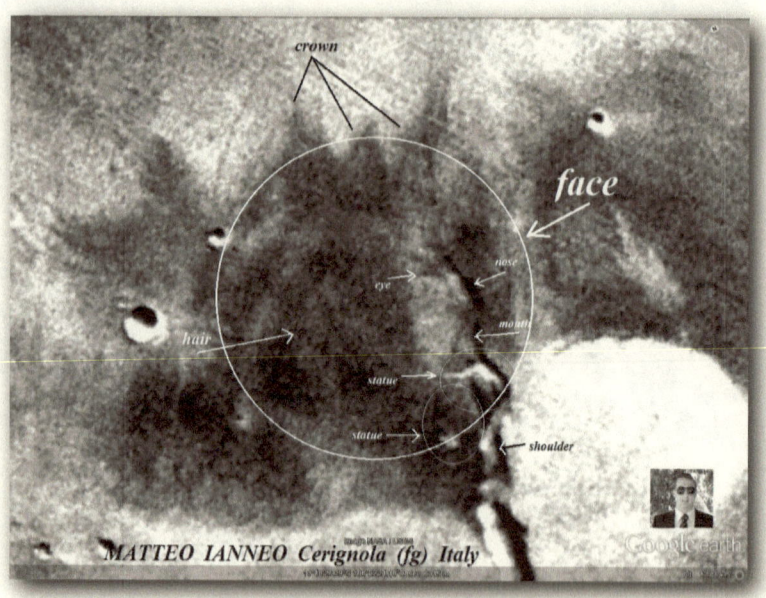

Image source: NASA / USGS

The profile of a king with a crown on his head.

In fact we can see a profile with an eye, nose like a parrot, and mouth with goatee beard. Below it we can see a standing statue, with black hair, and with an arm raised to the right as if pointing at something. The left arm rests at his side. Further along, in the centre at the bottom, we can see the profile of a statue with raised hair, like representations of gods such as Isis and others. We can also see his shoulder and arm. It seems like the pathway towards an entrance to somewhere, and we can see statues along the way.

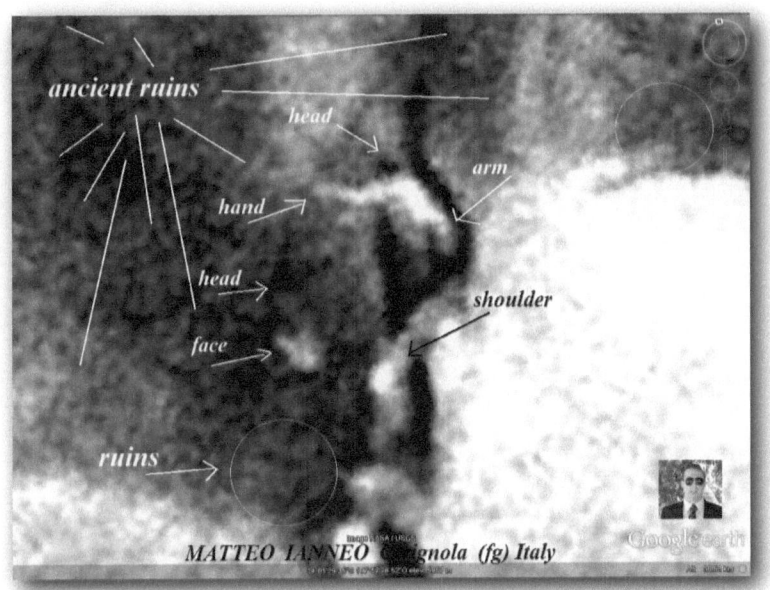

Image source: NASA / USGS

Here are the details of two statues, one small one above, with two arms – the right one raised with palm open, and below the profile of a female statue showing her face, raised hair, and with her shoulder and left arm visible.

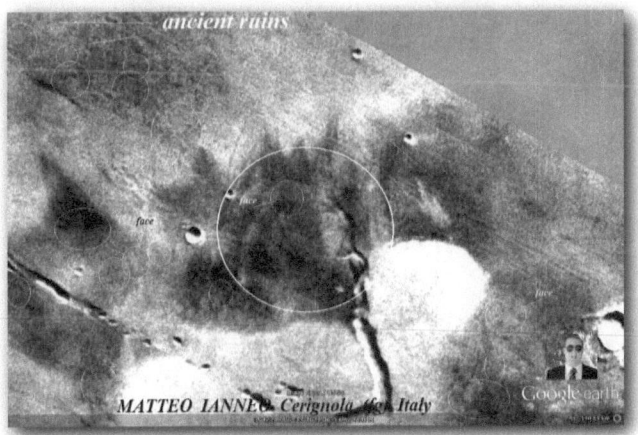

Image source: NASA / USGS

Another view.

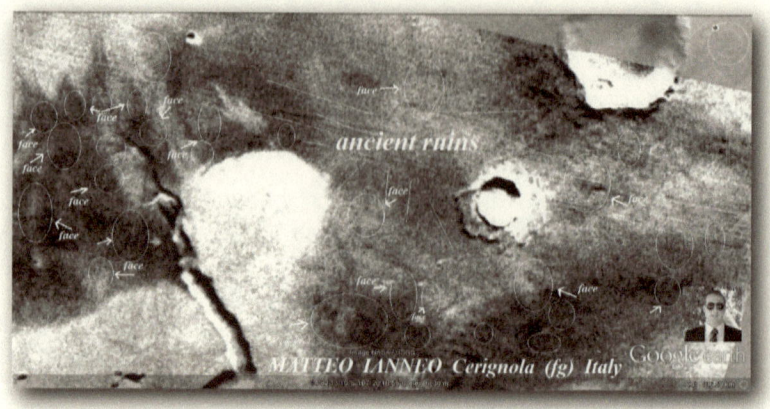

Image source: NASA / USGS

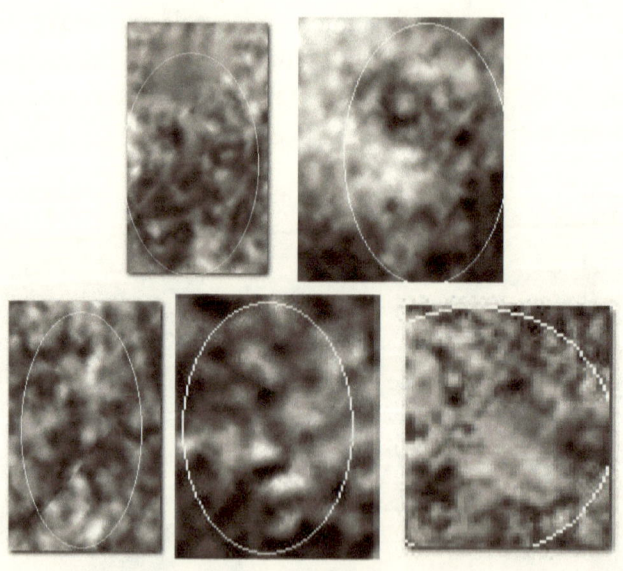

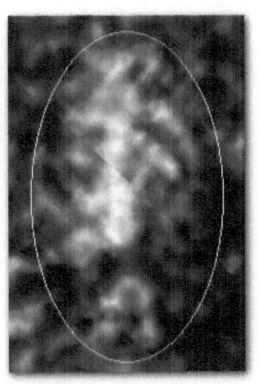

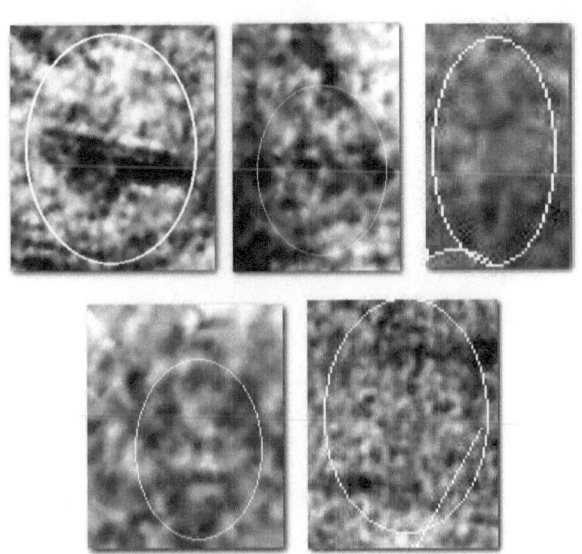

The position on
Google Earth is as follows:
Latitude 13°45'50.62"S   Longitude 107°59'19.87"W

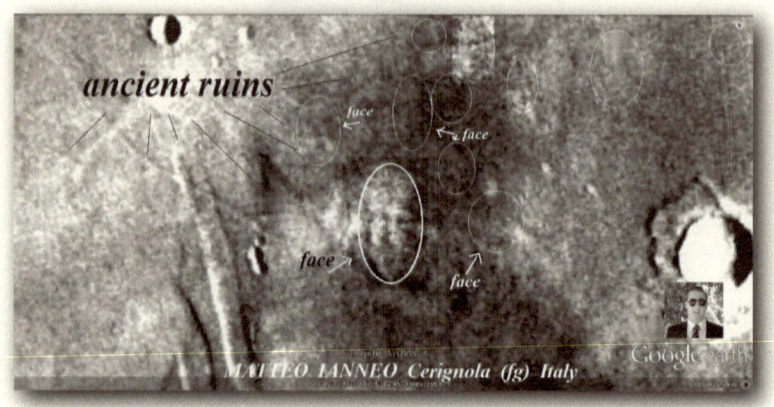

Image source: NASA / USGS

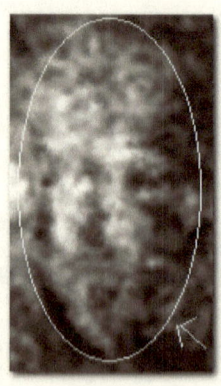

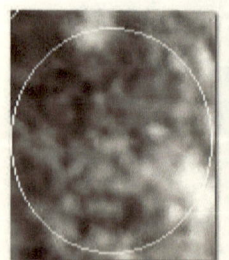
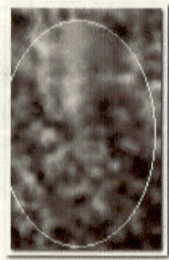
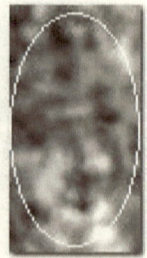

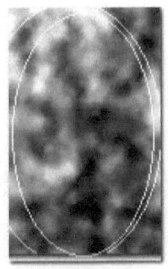 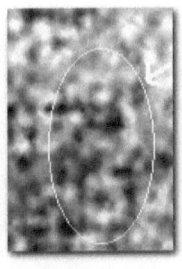 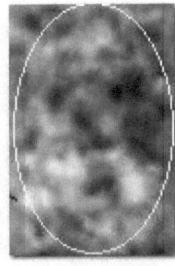

My eyes were ever more focused so I didn't miss anything. I looked carefully at what struck my senses. One day I found this element that at first glance did not give me any results, but, after some processing with filters, I found this detail, that of an ancient city.

The position on

Google Earth is as follows:

Latitude 36°11'27.97"N   Longitude 5° 23'17.17"W

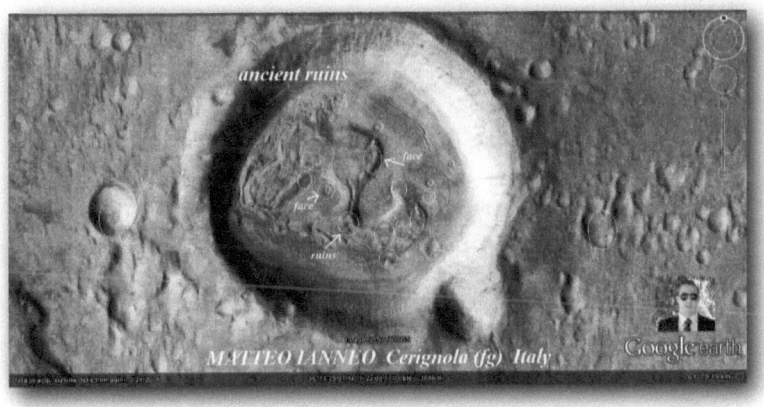

Image source: ESA/DLR/FU Berlin (G.Neukum)

We can see in this image its different elements. Monuments, at least those which remain, then, below, a type of entrance with a dome. It seems to be a temple with a representation that looks like a Russian pillbox. Of course, this is my hypothesis.

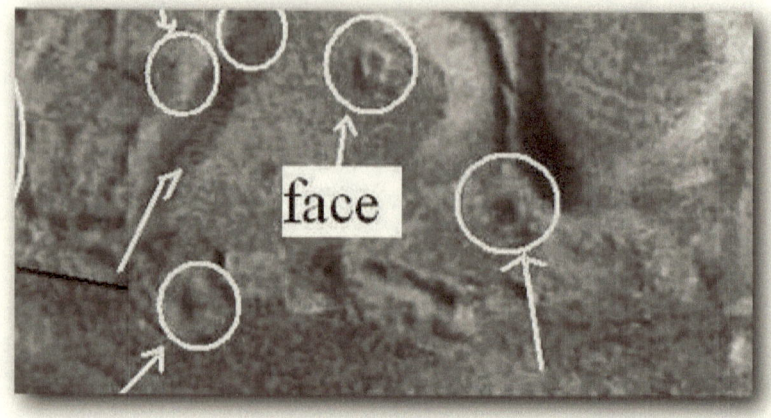

I zoomed in on the image to show you what I scrutinised. If I had not enlarged it many of you would not have noticed it. These are the remains of the ruins of an ancient civilisation. The majority of the ancient Martian civilisation is hidden inside large caves inside the mountains.

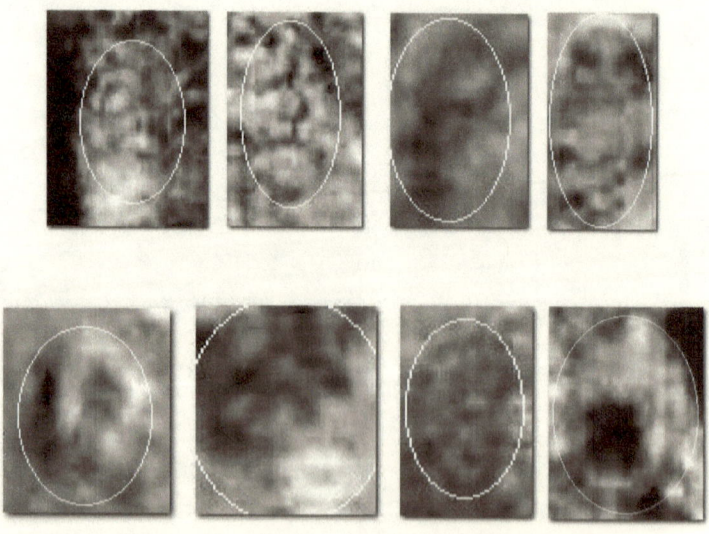

Image source: ESA/DLR/FU Berlin (G.Neukum)

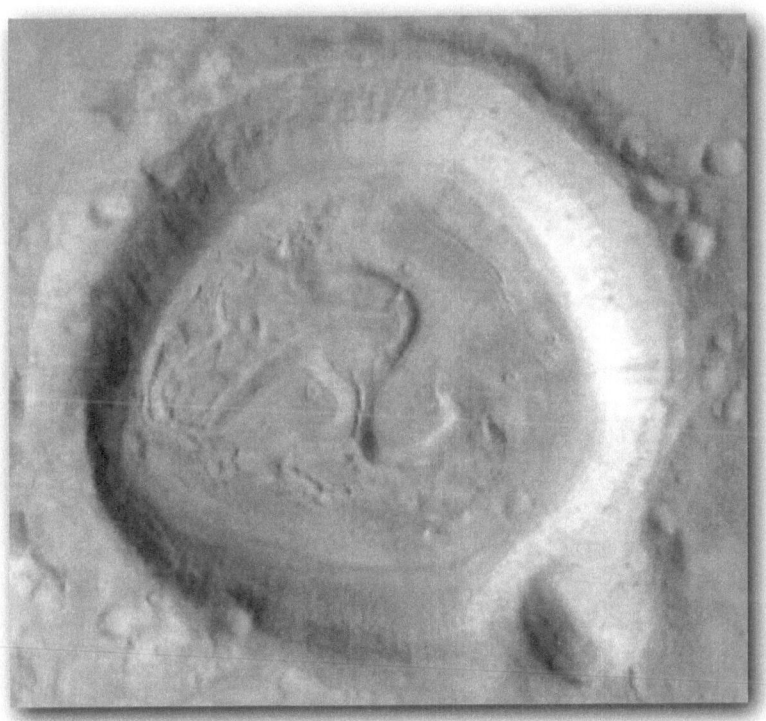

Image source: ESA/DLR/FU Berlin (G.Neukum)

The original.

Continuing my research, I discovered a familiar element: a representation of a face with a red hat on its head. I was beginning to wonder if indeed man had ever been on the Red Planet. This element, according to my analyses, depicts a man with a fur Busby on his head, like a Russian soldier. As well as the eye and nose, we can see a white beard and a collar. Although this is my hypothesis, I began to persuade myself that someone had been there for some time.

The position on

Google Earth is as follows:

Latitude 39°49'11.08"S   Longitude 139°52'7.18"W

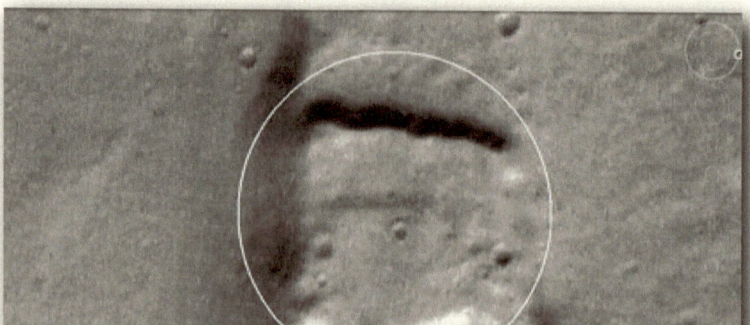

Image source: ESA/DLR/FU Berlin (G.Neukum)

Yes, the possibility that it is an optical illusion should not be excluded. However, if this was the result of natural events, we would have to admit that we are facing a very intelligent nature indeed.

Image source: ESA/DLR/FU Berlin (G.Neukum)

With filter.

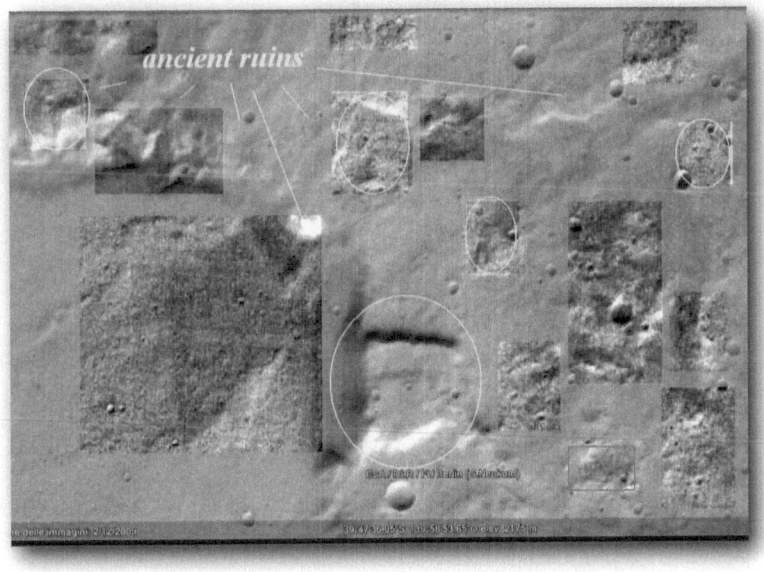

Image source: ESA/DLR/FU Berlin (G.Neukum)

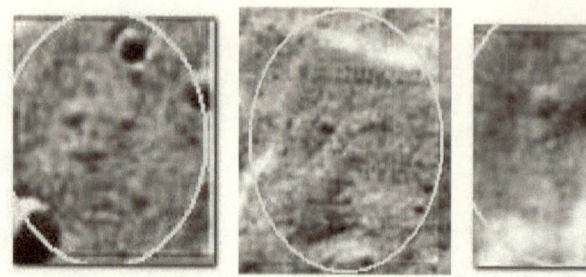

The position on
Google Earth is as follows:
Latitude 39°54'45.81"S   Longitude 139° 5'58.30"W

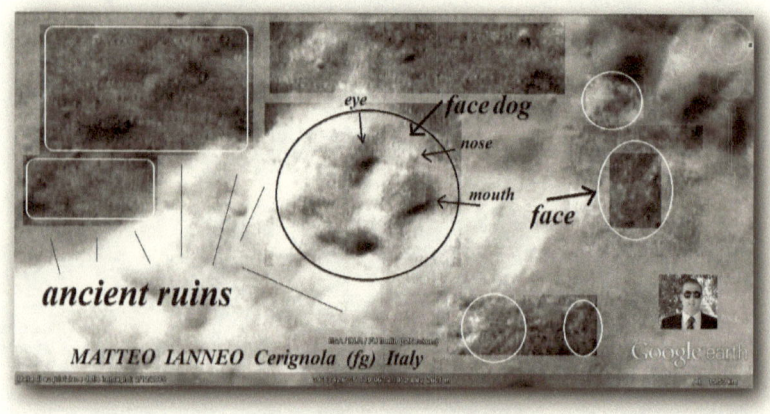

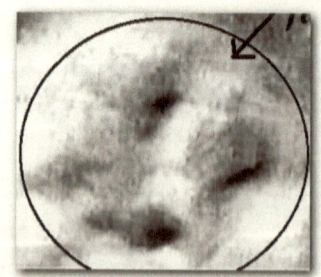

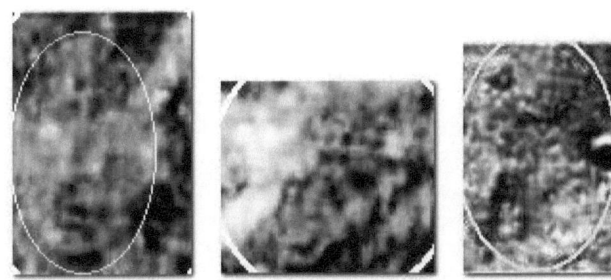

Image source: ESA/DLR/FU Berlin (G.Neukum)

The original.

This face shows something different: the face of an alien being.

The position on

Google Earth is as follows:

Latitude 37°27'14.20"N   Longitude 4°36'11.18"E

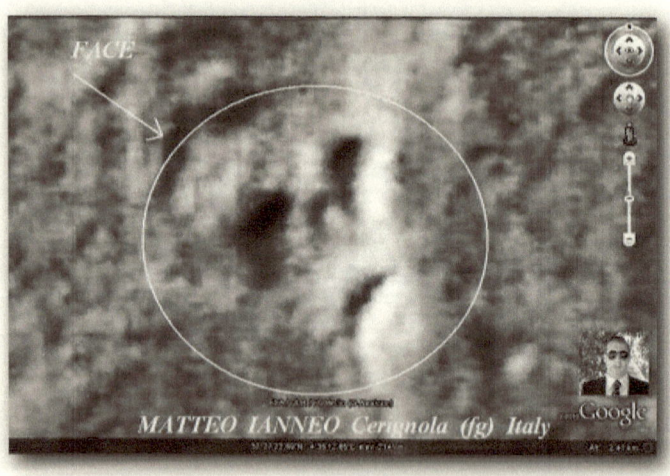

Image source: ESA/DLR/FU Berlin (G.Neukum)

Eyes, mouth, nose – elements that are present every day in our common senses. If this face could have been a coincidence of nature, what me made me change my mind is the detail in the next image.

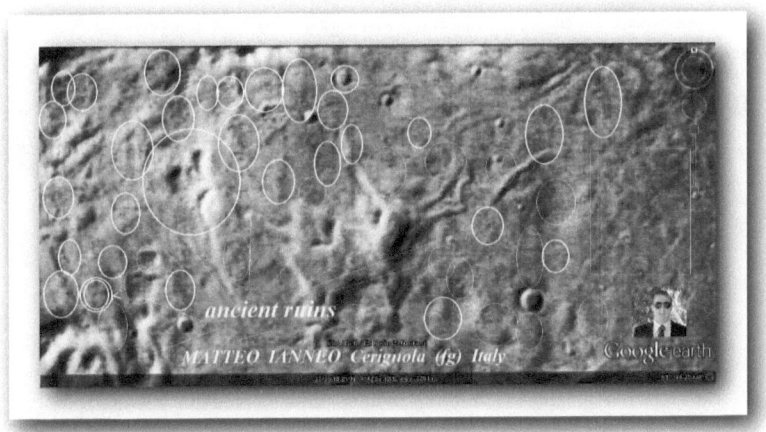

If you look carefully at the item to the left of the image, highlighted by a circle, you can see a woman's face looking down - another monument. Other details can be seen in the next image.

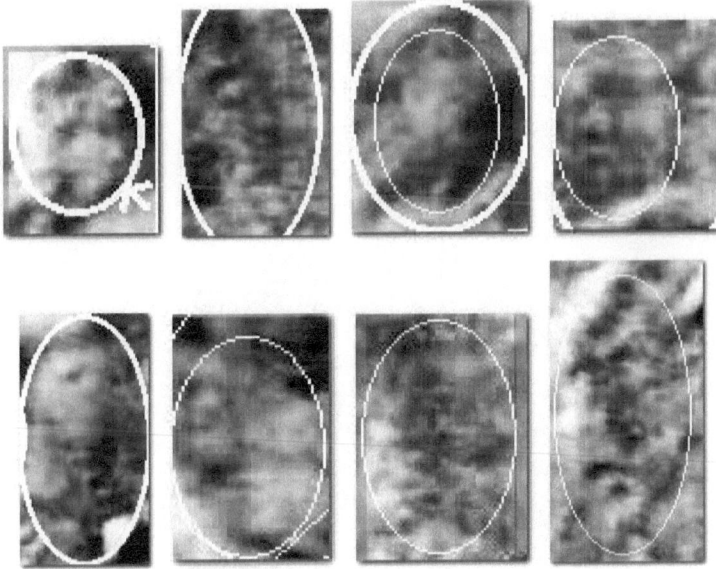

immagine: ESA/DLR/FU Berlin (G.Neukum)

To the top right a face which can be reached by stairs can be seen. These could be ancient ruins which have deteriorated over time.

 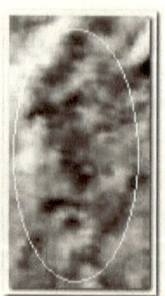 

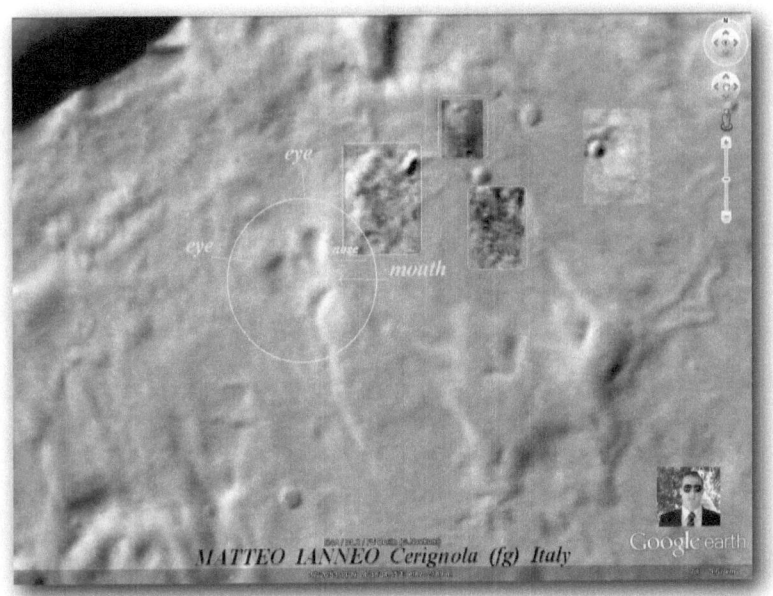

immagine: ESA/DLR/FU Berlin (G.Neukum)

The original.

Another interesting face in profile.

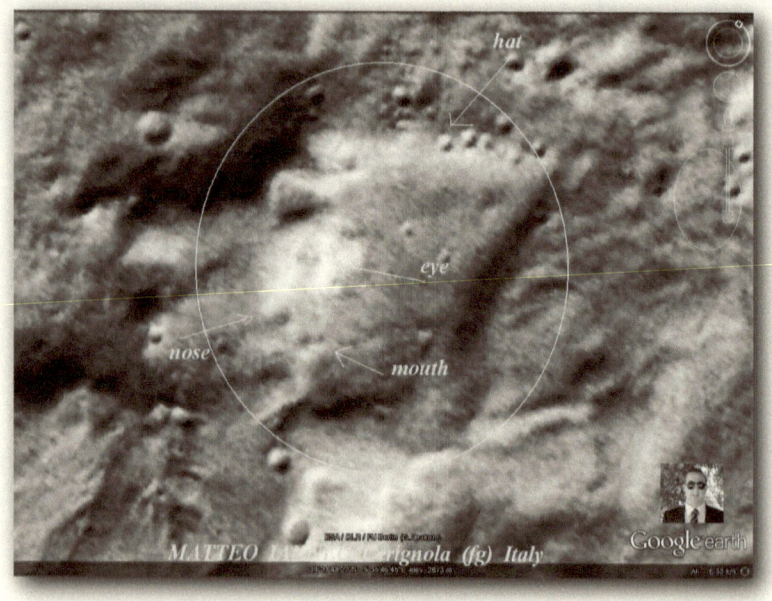

Image source: ESA/DLR/FU Berlin (G.Neukum)

Another face with something on its head, with quite a long beard, similar to the one worn by Egyptians.

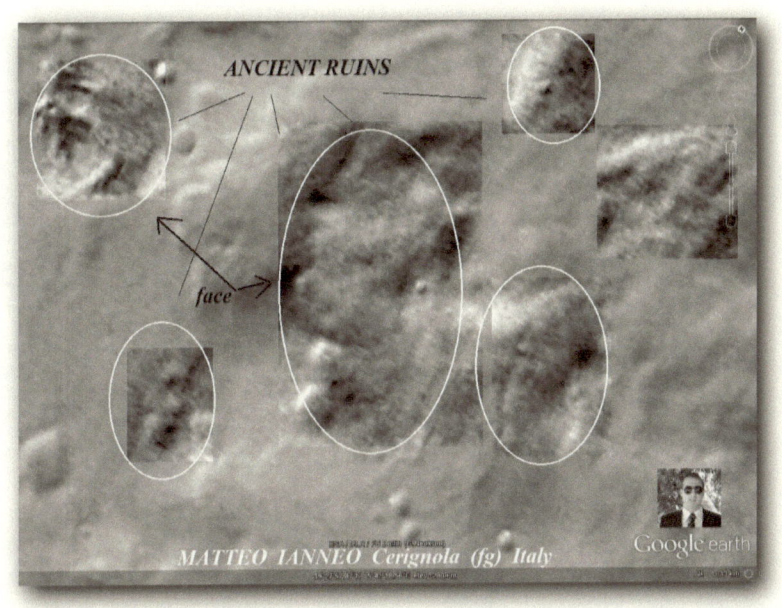

Image source: ESA/DLR/FU Berlin (G.Neukum)

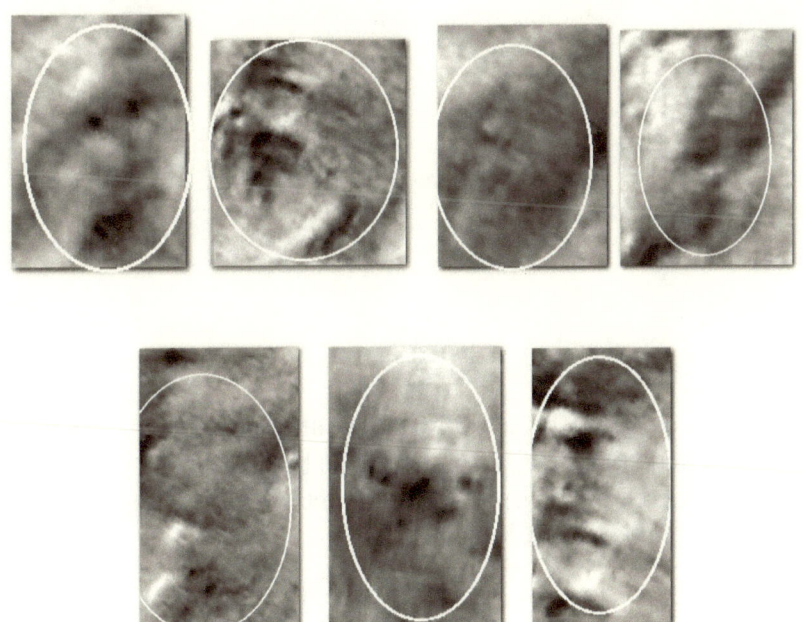

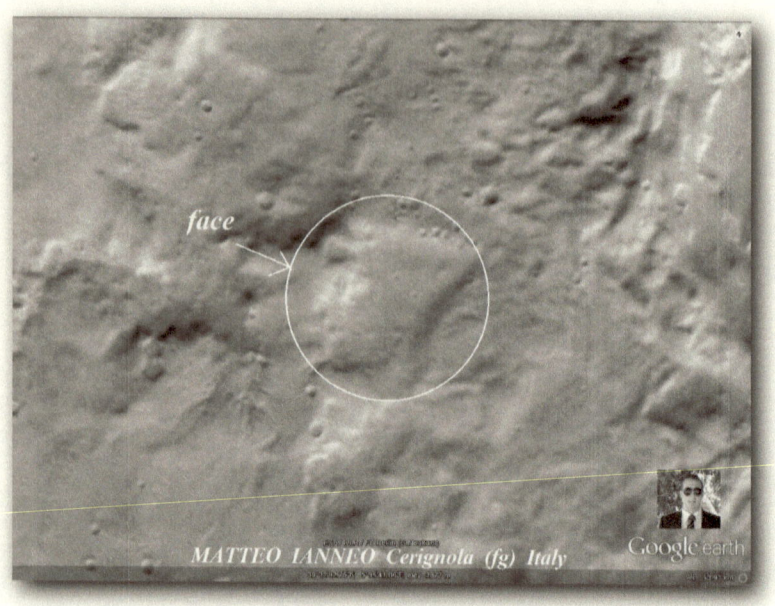

Image source: ESA/DLR/FU Berlin (G.Neukum)

The original.

As I looked and wandered with an attentive eye in order to capture interesting details, I noticed this being with legs and feet - a being carved in stone as a testament to a civilisation which existed at the time. Just look at it!

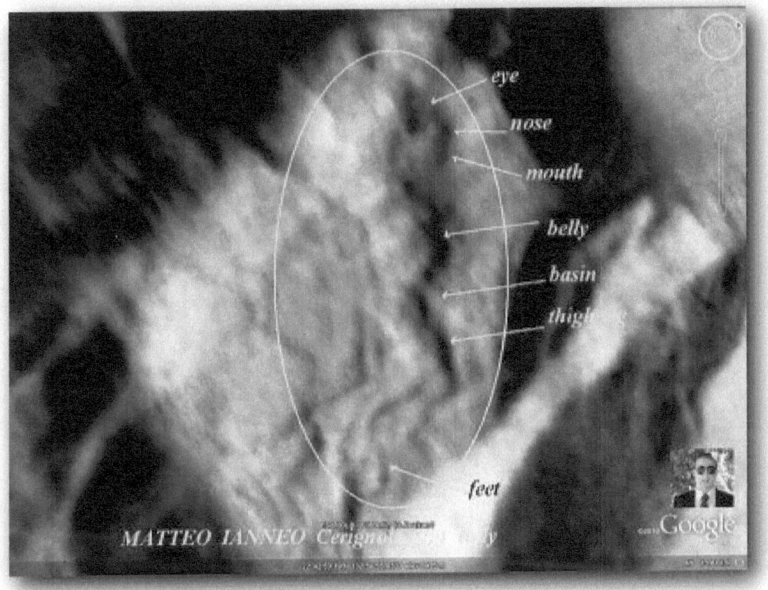

Image source: ESA/DLR/FU Berlin (G.Neukum)

If we analyse it together, we see the face of a being with an eye, nose, mouth, stomach and pelvis, thighs and feet – a curious being in all respects, a dominant being of that civilisation.

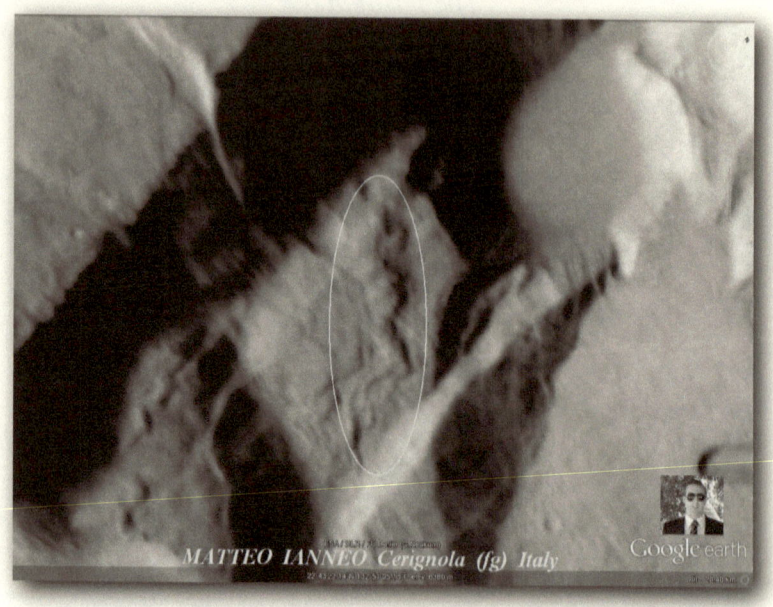

Image source: ESA/DLR/FU Berlin (G.Neukum)

Here is the original view.

Yet another detail: a being with a very large mouth, again in profile.

The position on
Google Earth is as follows:
Latitude 13°54'39.89"S   Longitude 139°50'17.65"W

Image source: NASA / USGS

A kind of statue with a mouth like a fish. You can glimpse the eye, nose and mouth. It looks like a monument from whose fissures you could access the city, hidden inside the mountain.

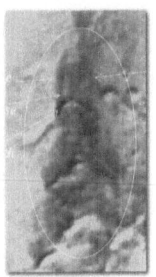

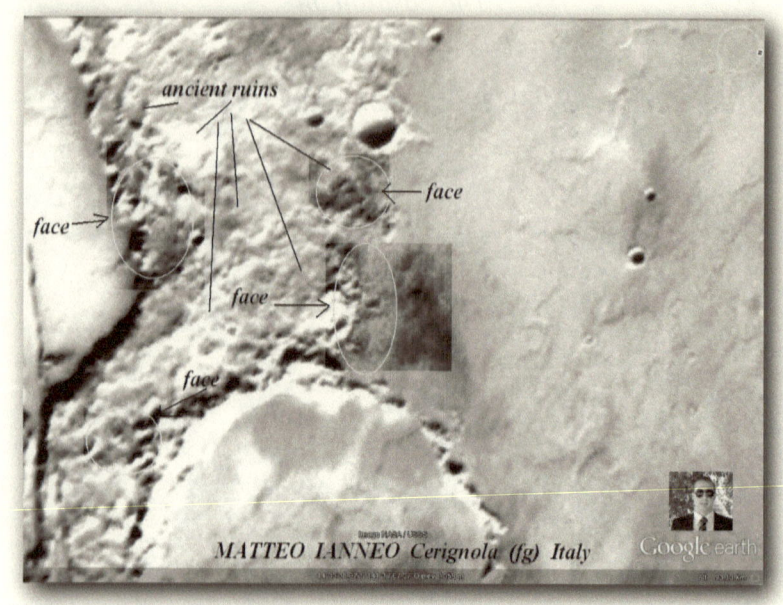

Image source: NASA / USGS

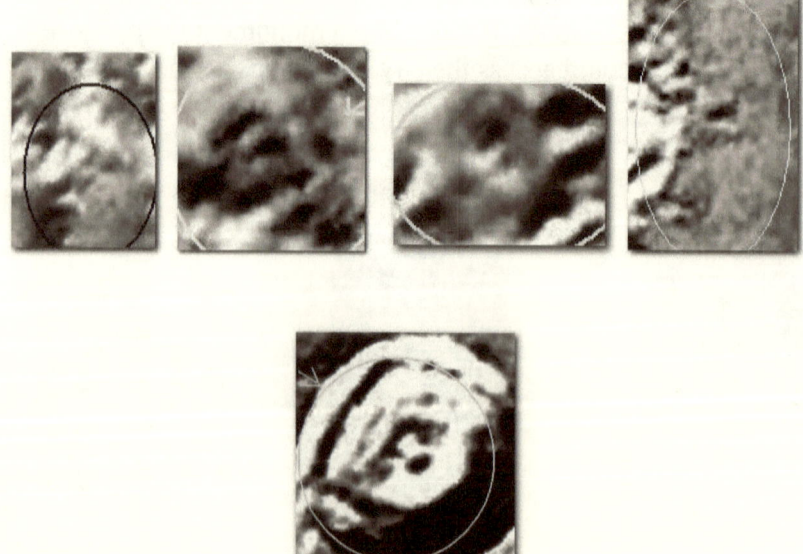

MATTEO IANNEO Cerignola (fg) Italy

Image source: NASA / USGS

The original.

My exploration continued in an attempt to find other precious details. I identified many faces, but I discarded from my exploration those which were not too distinguishable or rather inconspicuous.

The following struck me most and I decided to keep them in my explorer's bag.

The position on

Google Earth is as follows:

Latitude 37°2'54.94"N   Longitude 12°13'4.19"W

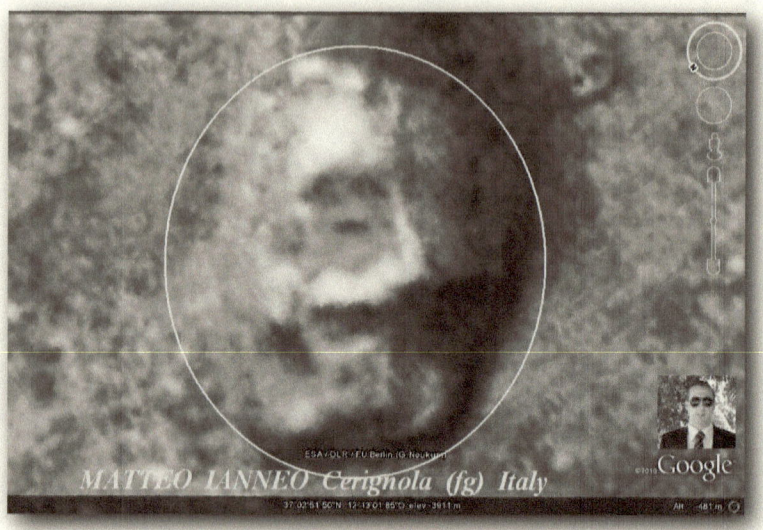

Image source: ESA/DLR/FU Berlin (G.Neukum)

With my imagination, looking at this face, I immediately thought of a leader, maybe a man who fought in defence of his people and his city. We can see the closed eye clearly.

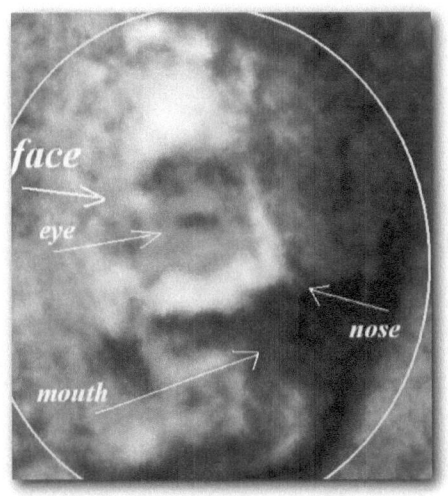

Here, the details: eye, nose, mouth and perhaps a protective helmet worn for protection.

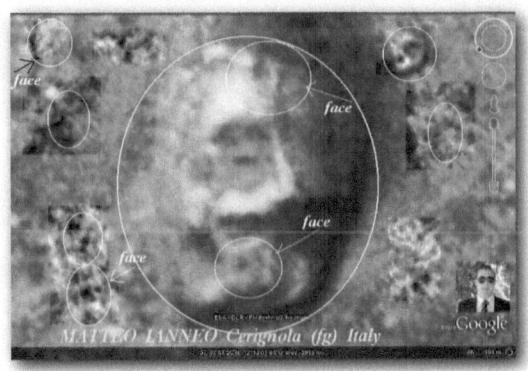

Image source: ESA/DLR/FU Berlin (G.Neukum)

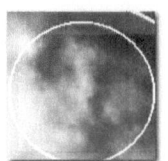 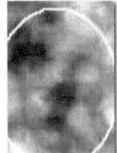 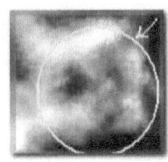 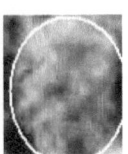

Since there are always surprises, here is a new animal which has become part of my collection.

The position on
Google Earth is as follows:
Latitude 37°10'25.36"N   Longitude 4° 4'48.48"E

Image source: ESA/DLR/FU Berlin (G.Neukum)

At first glance, what hit my retina and my senses was the shape of an ostrich - the long neck, beak and an eye. This discovery reinforced the certainty in me that on the Red Planet there are many similarities to Earth: life born simultaneously on both planets, or a migration or escape from Mars to Earth.

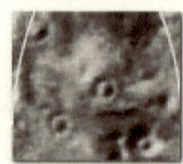

Image source: ESA/DLR/FU Berlin (G.Neukum)

In detail.

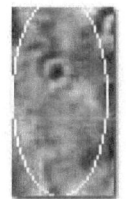 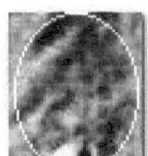  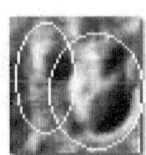  

Adding to the section dedicated to animals, I found this set of elements.

The position on

Google Earth is as follows:

Latitude 36°22'47.39"N   Longitude 22°58'41.07"E

Image source: ESA/DLR/FU Berlin (G.Neukum)

Below, at the bottom, there is an example of a canine specimen which resembles a cartoon; above it there is another animal with mouth, closed eyes and elongated neck. Further above I found the figure of a pharaoh in profile, isolated and highlighted below.

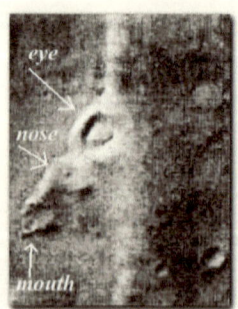
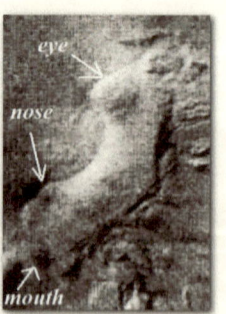

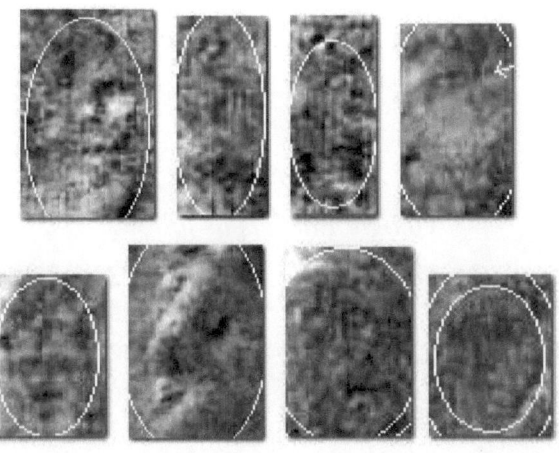

Image source: ESA/DLR/FU Berlin (G.Neukum)

Here in another position.

I've made the pharaoh free and, by dissociation, I've highlighted him.

The position on
Google Earth is as follows:
Latitude 36°17'39.44"N   Longitude 22°58'29.99"E

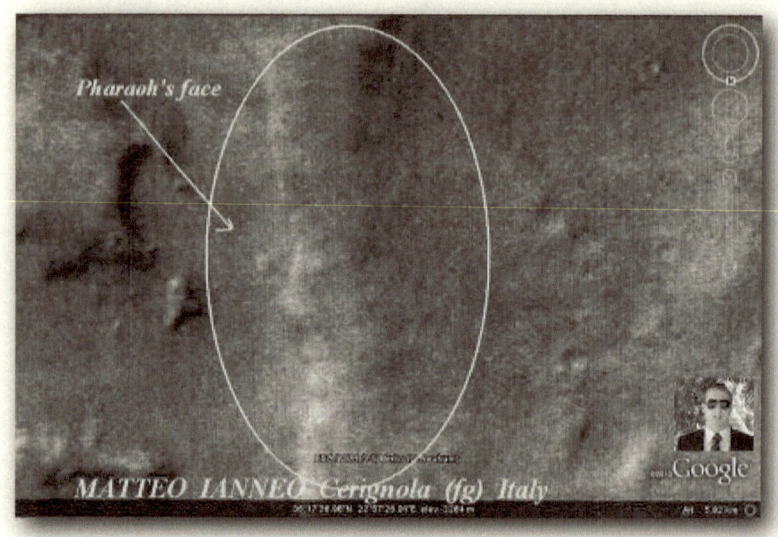

Image source: ESA/DLR/FU Berlin (G.Neukum)

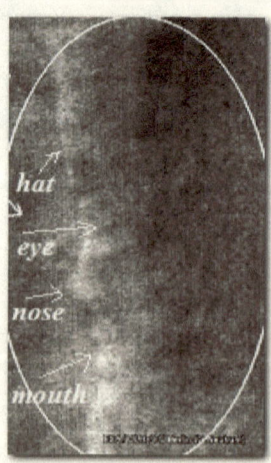

You need to have a particularly attentive eye. Some of you will think I am crazy. In any case, I see a pharaoh. I see a profile with pronounced chin, mouth, nose, eye and forehead. On his head he has a headdress, typical of the pharaohs.

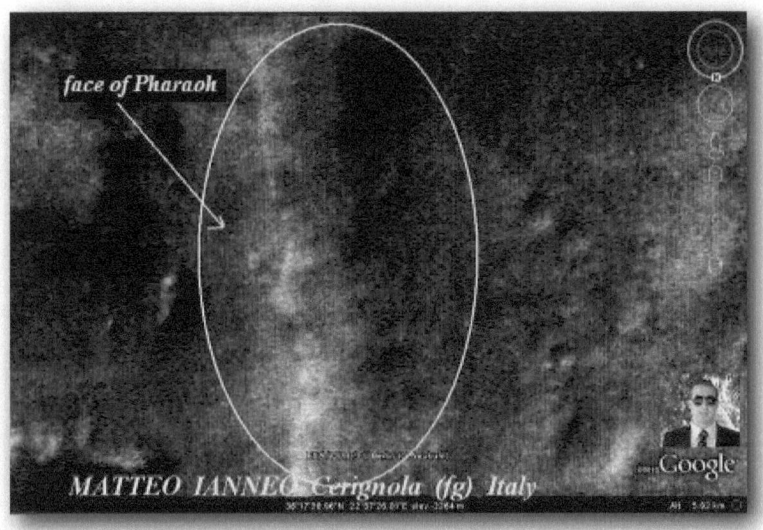

Image source: ESA/DLR/FU Berlin (G.Neukum)

Here it is treated with a filter which highlights the details. I believe that these sculptures are so old as to have deteriorated through erosion and the passing of time.

I have also found many other pharaohs.

The position on
Google Earth is as follows:
Latitude 72°39'31.45"N   Longitude 124°12'34.53"W

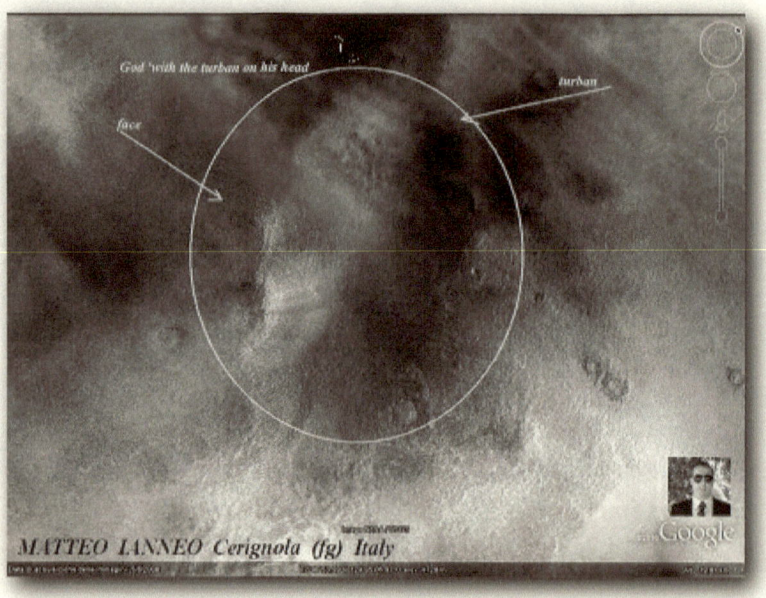

Image source: NASA / USGS

Here it is. Unfortunately it was banned perhaps by those who do not want it to be discovered. Note the head wrapped in a turban, the hidden face where you can catch a glimpse of the nose, eye and mouth, with a long beard typical of the pharaohs. Someone has tried to hide the truth... but we can observe it.

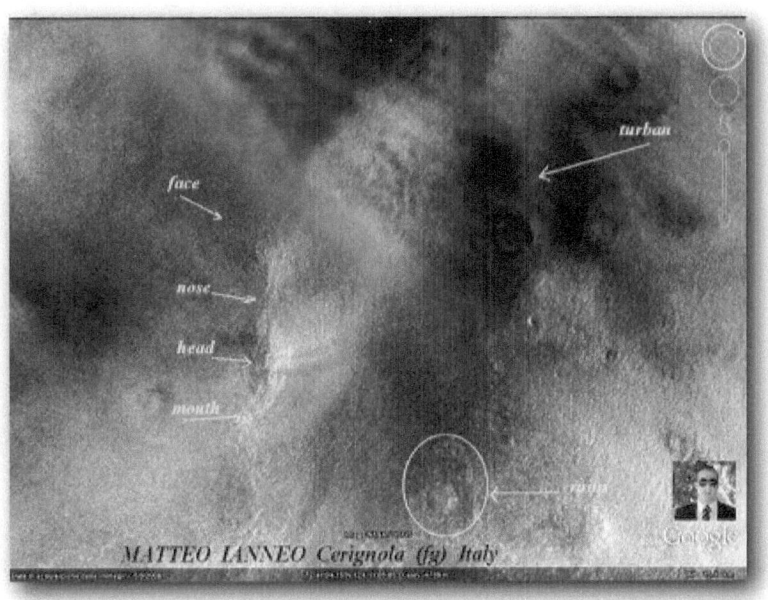

Image source: NASA / USGS

Here is our pharaoh.

Beneath him, in the centre at the bottom, you can glimpse an ancient structure like a colonnade, perhaps used as the entrance to another city. I have no evidence as to how they could build these huge faces, but I have some ideas. Someone used tools from above. Spaceships able to create the sculptures without help from the people. Using computers they modelled faces related to their history. Somewhere there was an access to embark and disembark the spaceships, while life developed beneath the surface of the planet. I am convinced that if we were to explore Mars beneath the surface, it would take centuries to understand and learn about the civilisations that have populated it and its dynasties.

Obviously this is just my hypothesis. Now follow me carefully.

A new creature.

The position on

Google Earth is as follows:

Latitude 11°10'13.46"S   Longitude 121° 7'28.88"W

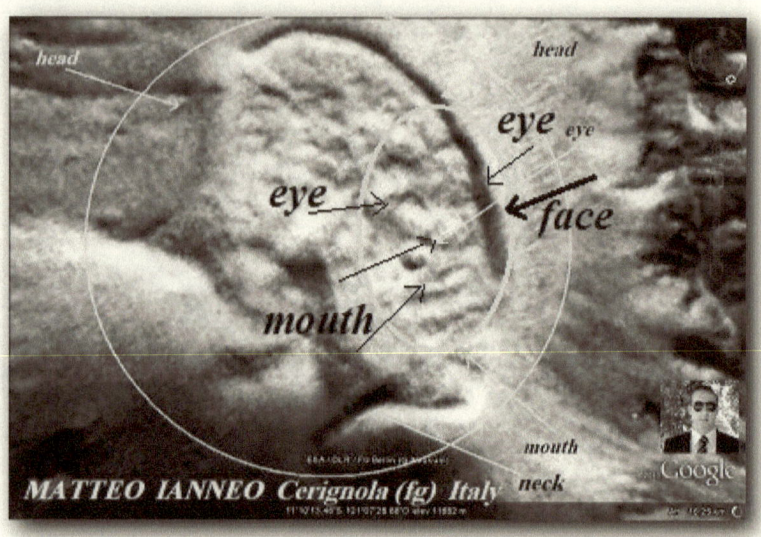

Image source: ESA/DLR/FU Berlin (G.Neukum)

In this image you can see a creature which is wearing something like a helmet. From its form, notice its mouth which is similar to that of a fish. I imagine a creature from the abyss. I've called it *The man from the abyss*. You can see an eye, the head and neck.

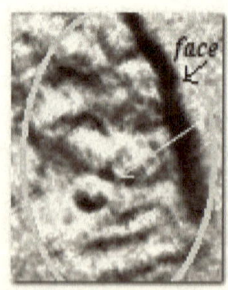

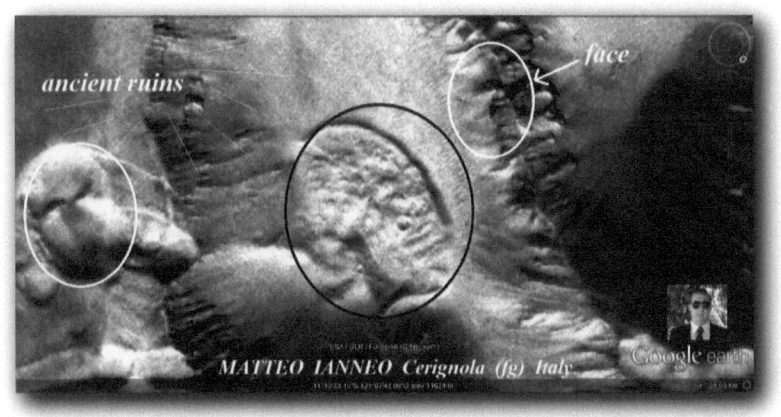

ancient ruins

face

MATTEO IANNEO Cerignola (fg) Italy

Google earth

Image source: ESA/DLR/FU Berlin (G.Neukum)

With another filter.

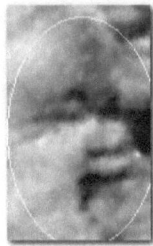

Continuing to scan the surface of Mars I found another example of an animal, perhaps a wolf or a bear.

The position on

Google Earth is as follows:

Latitude 14°4'57.83"S   Longitude 133° 7'57.31"E

Image source: NASA / USGS

I wanted to keep this image. Even if it were to be found to be pareidollia, what bothers me is the object placed in the top centre. It has the shape of a nutcracker. I don't believe that nature has shaped this object for our eyes.

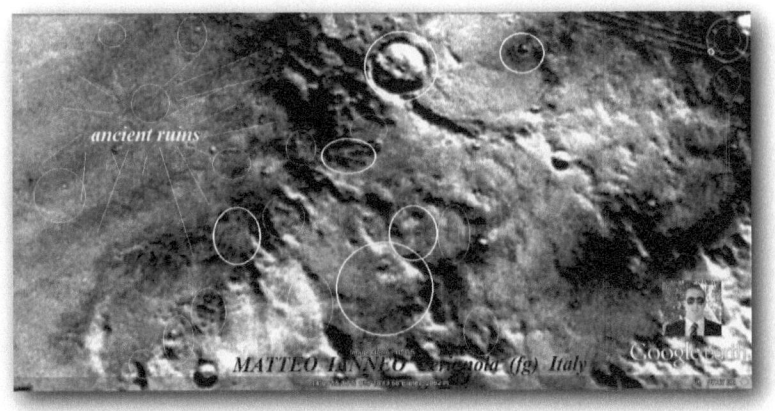

Image source: NASA / USGS

Here in detail.

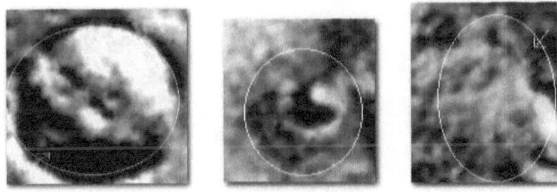

I found this face, shall we say extraterrestrial?

The position on

Google Earth is as follows:

Latitude 29°16'47.08"N   Longitude 37°22'37.04"W

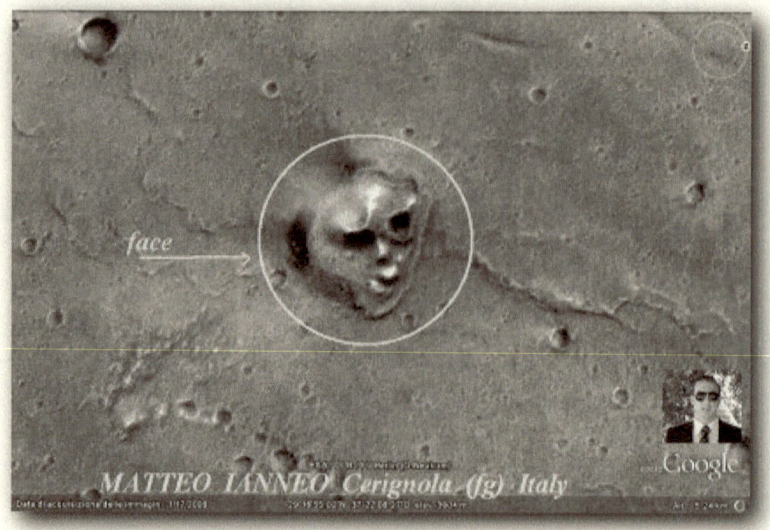

Image source: ESA/DLR/FU Berlin (G.Neukum)

Here too we see a face. I don't believe that all of this can be dismissed as an optical illusion. In the next frame you can see the face, two eyes, mouth and nose.

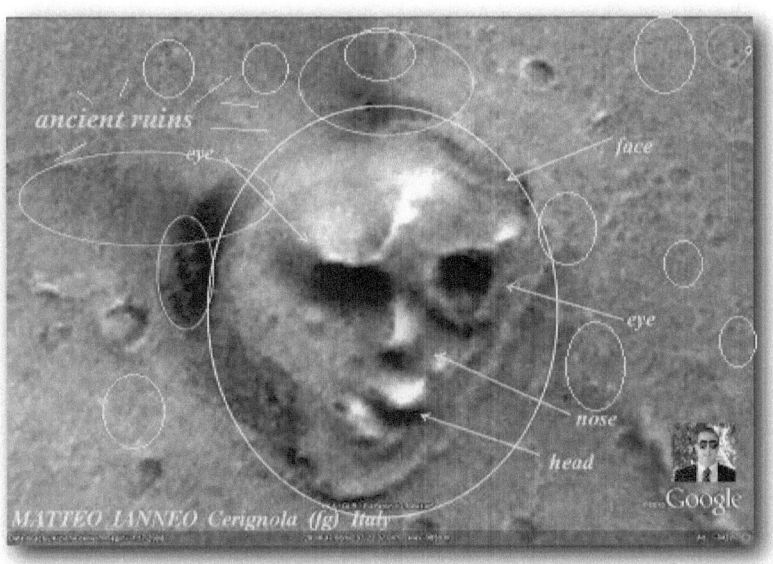

Image source: ESA/DLR/FU Berlin (G.Neukum)

Perhaps here I wanted to exaggerate, but if you look to the left of the face there is a type of erect sculpture. On his head, to the top centre, you can see a small face and parts of a wall, even if almost demolished, visible also to the left. My imagination leads me to think that those eyes are the entrance and exist for the aircraft that have their base hidden beneath the ground.

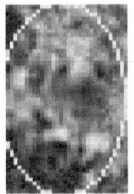

In the photo below we can see the profile of a being which also has something on its head.

The position on

Google Earth is as follows:

Latitude 35°22'30.06"N   Longitude 131°42'15.09"E

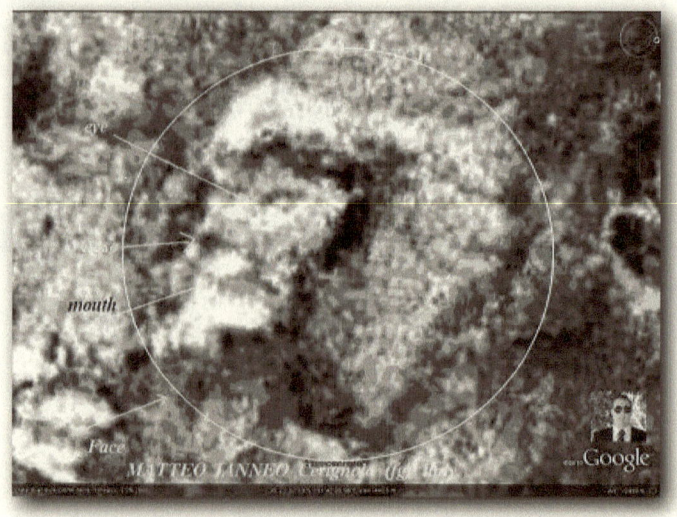

Image source: NASA / USGS

I know that you are asking: "Yet another face?" Of course!

These faces hide their ruined cities, destroyed by something we will never know. Look closely and you will see small significant details. I hope you can see at least one.

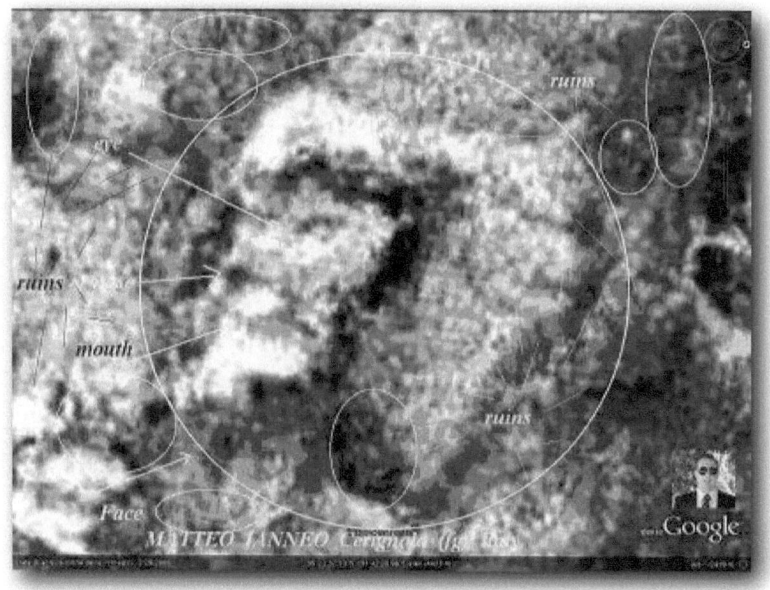

Image source: NASA / USGS

Image source: NASA / USGS

If you concentrate you will see various details of ancient buildings long since abandoned. These are the ruins of an ancient civilisation. If we removed our seas, we would also find more history here on Earth.

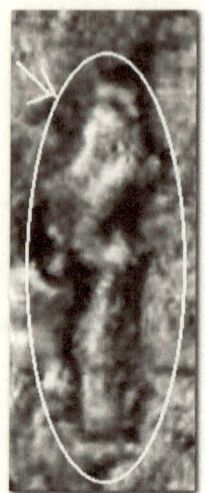

A statue.

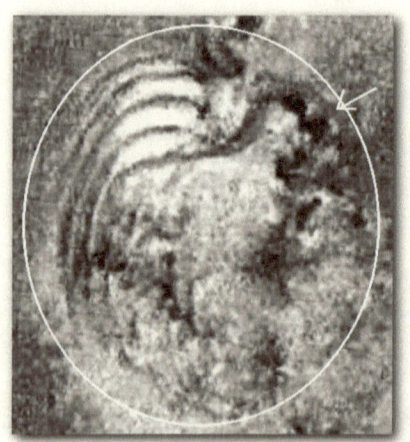

A cockerel.

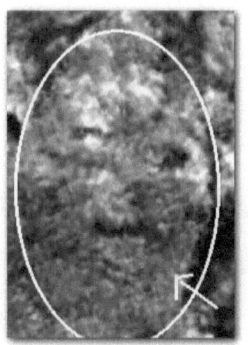

A face.

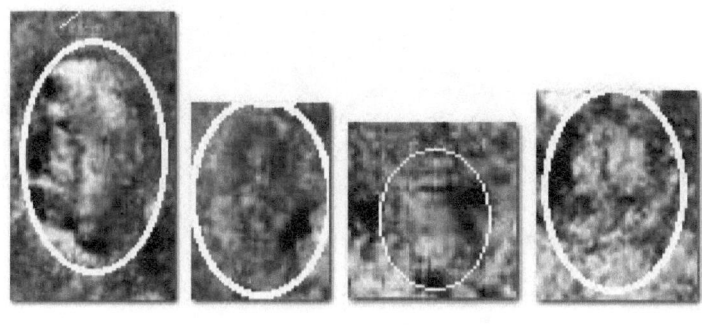

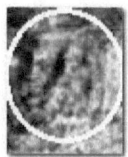

Let's leave these details and move on to another that represents something very ancient.

The position on
Google Earth is as follows:
Latitude 40° 9'27.96"N   Longitude 39°27'30.08"E

Image source: ESA/DLR/FU Berlin (G.Neukum)

A carved rock depicting a being with clearly visible characteristics: eye, long nose and mouth.

Image source: ESA/DLR/FU Berlin (G.Neukum)

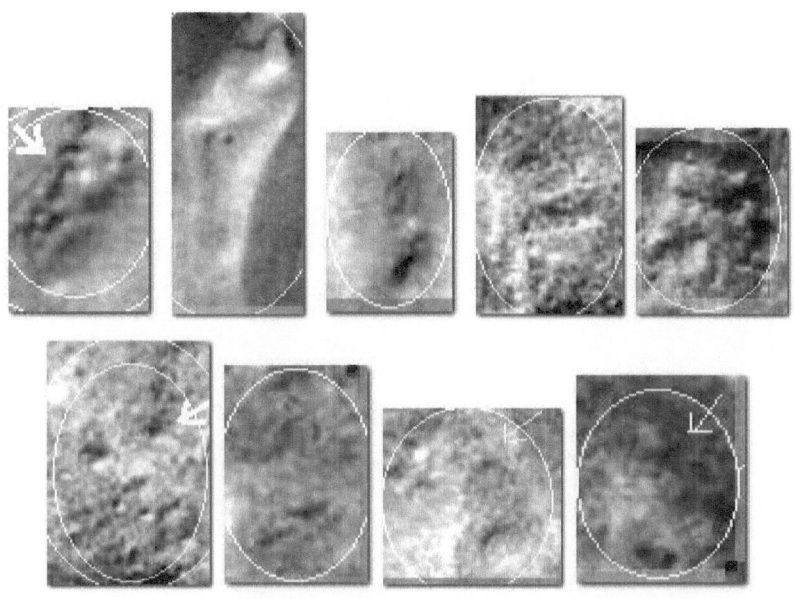

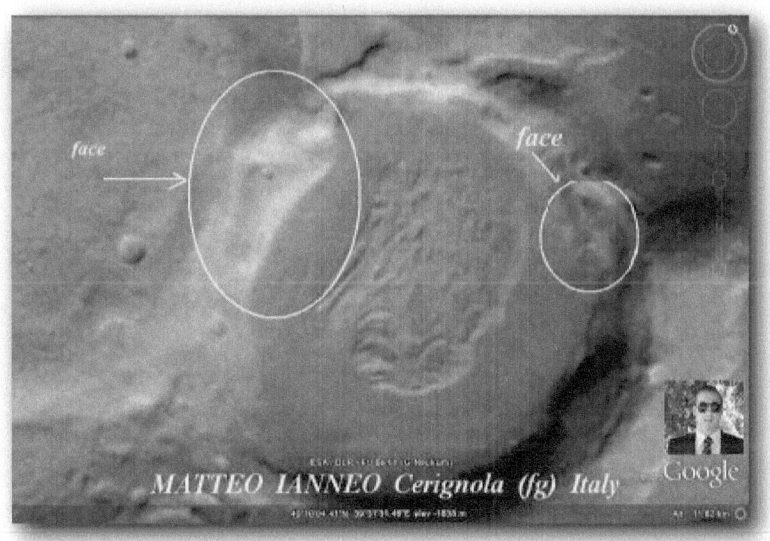

Image source: ESA/DLR/FU Berlin (G.Neukum)

Here, the original.

To the right of the main subject there is another face.

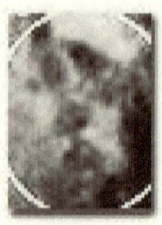

Now here is a canine element: a dog's head.

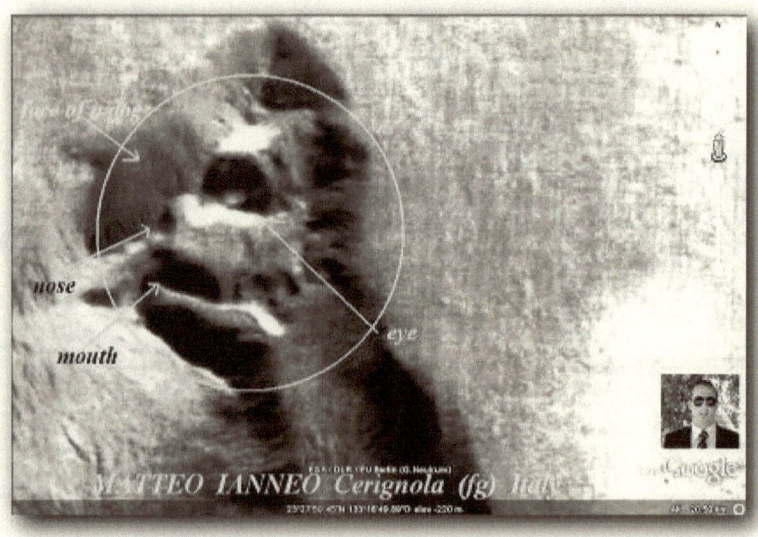

Image source: ESA/DLR/FU Berlin (G.Neukum)

This image, in its turn, is the entrance to an underground area, the entrance to a city which could be located in the mouth of the subject. This is a common feature to all the people who lived on Mars: the construction of underground cities accessed through the mouth of their god. In this case the mouth of a dog.

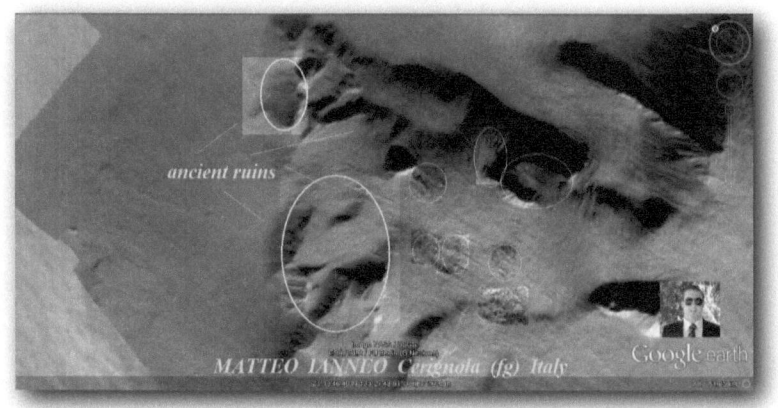

Image source: ESA/DLR/FU Berlin (G.Neukum)

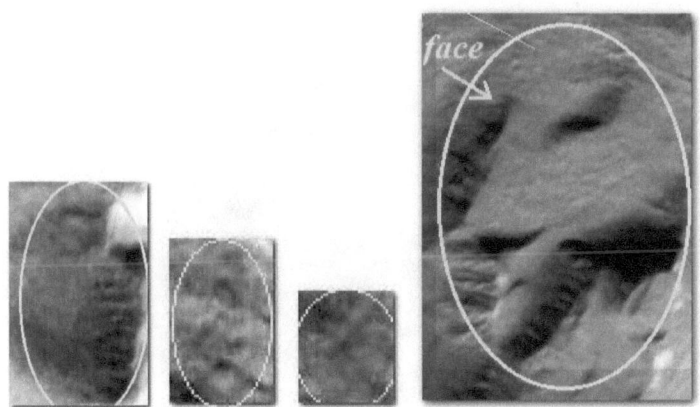

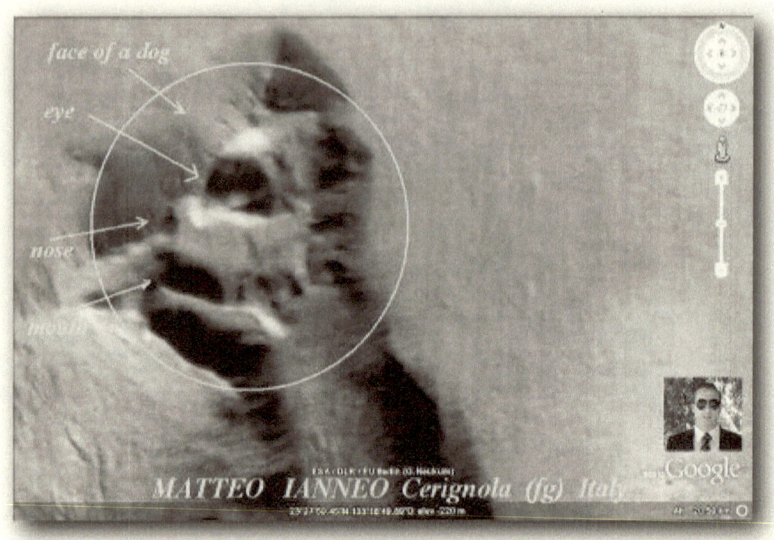

Image source: ESA/DLR/FU Berlin (G.Neukum)

The original.

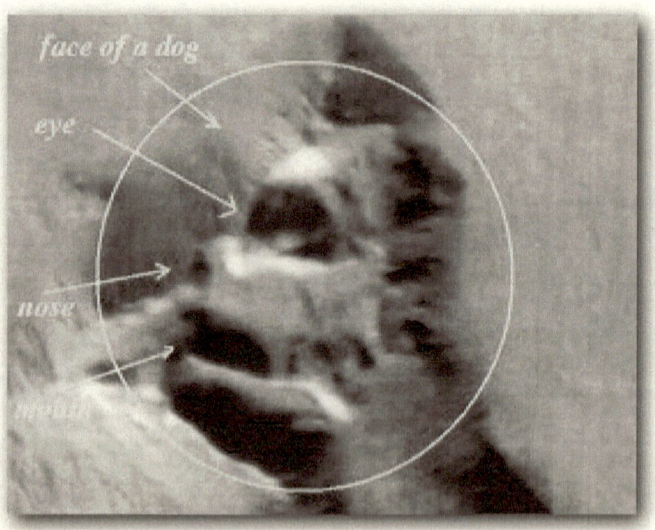

I found it difficult to attribute a similarity or association of ideas to this image, but after careful consideration, I realised that it is a woman's face.

The position on

Google Earth is as follows:

Latitude 18°11'11.39"N   Longitude 43°59'59.76"W

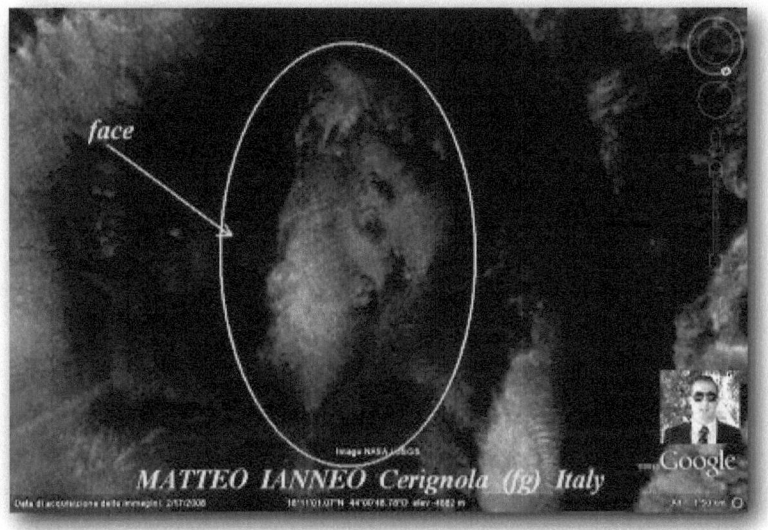

Image source: NASA / USGS

A goddess who has been waiting for her god for some time. Saddened by the loss of her family, she remained here to wait for someone to take her away. Look at her eyes: you can even see her pupils.

In the next frame, there are more details.

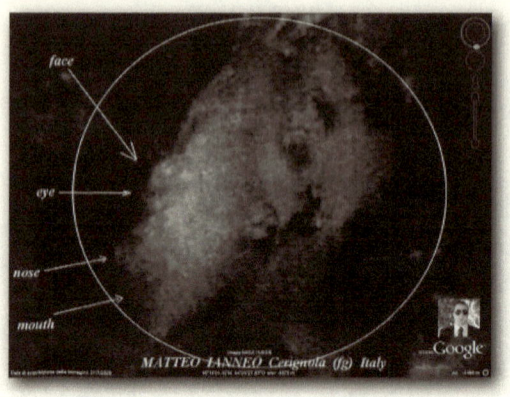

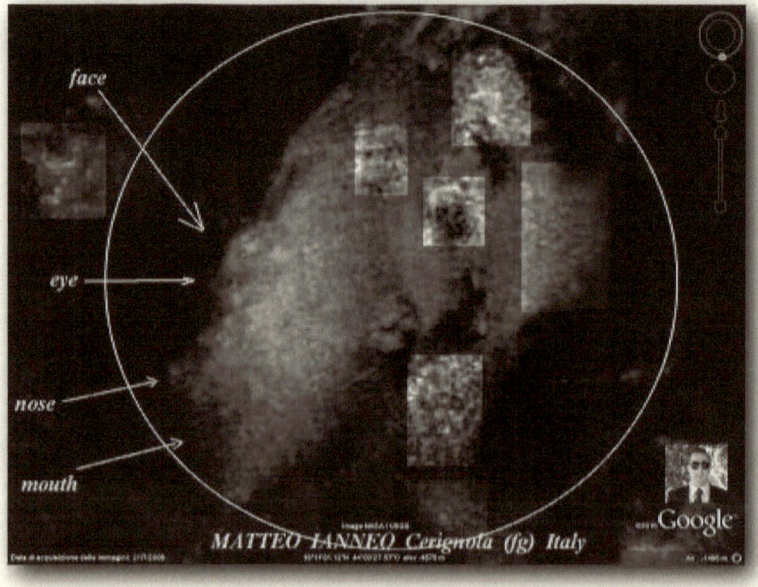

Image source: NASA / USGS

Here is the profile.

You can see the eye with the pupil, nose with nostrils, mouth and chin. A face concealed in a place where no one would ever be able to reach.

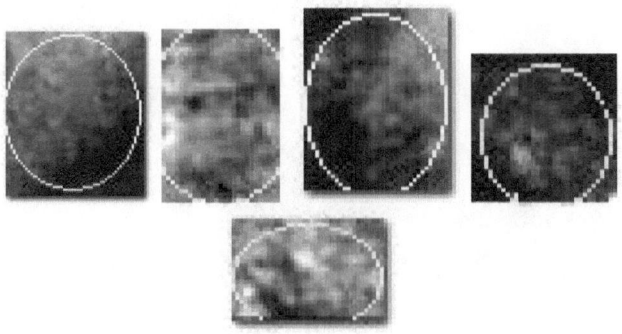

Finally something different, a dragon.

The position on

Google Earth is as follows:

Latitude 18°19'10.32"N   Longitude 44° 1'51.52"W

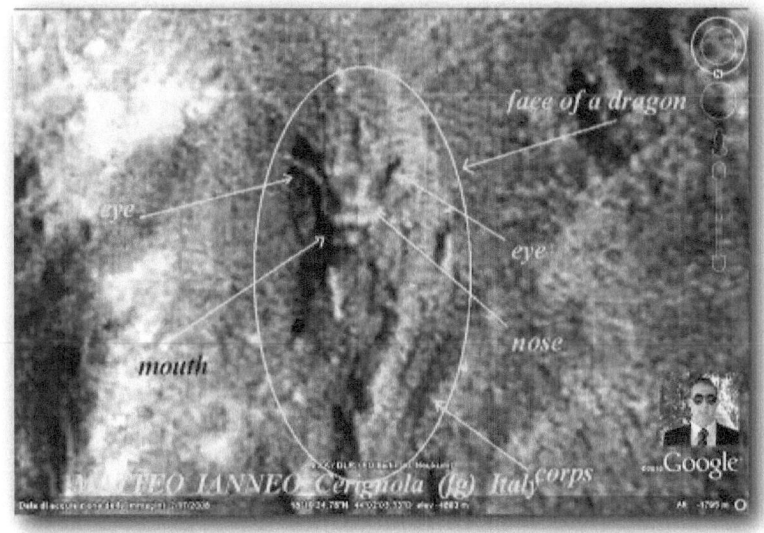

Image source: ESA/DLR/FU Berlin (G.Neukum)

I couldn't believe my eyes. Slit eyes, mouth typical of reptiles and the body of a snake - a beautiful Chinese dragon.

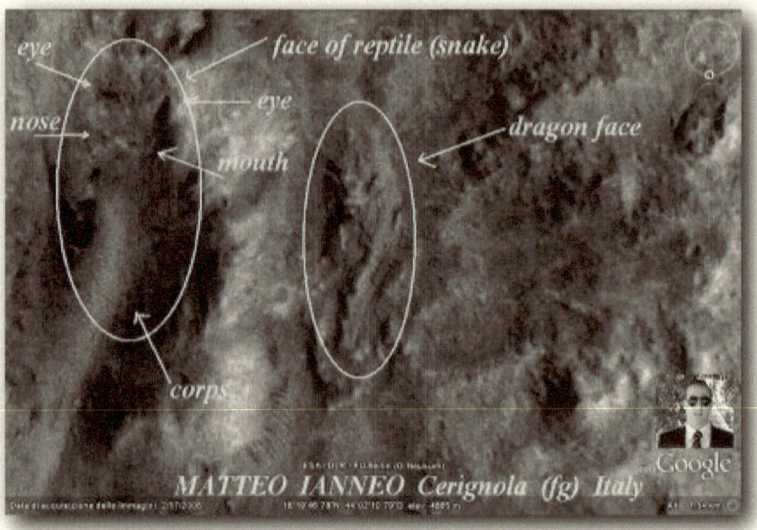

Image source: ESA/DLR/FU Berlin (G.Neukum)

Here it is, accompanied by another reptile – it looks like a snake – to the left of the image. I'm not a historian, but looking at this picture I think there is some connection with Earth.

Image source: ESA/DLR/FU Berlin (G.Neukum)

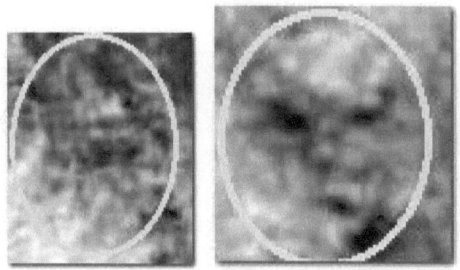

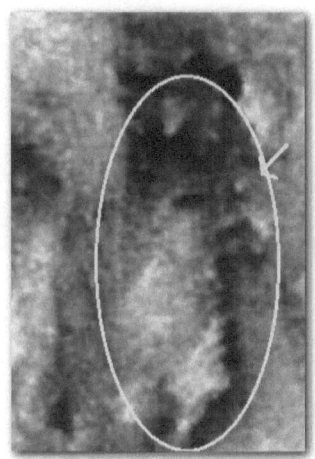

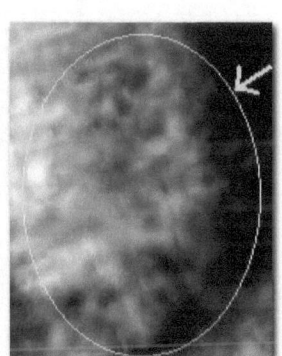

Yet another…

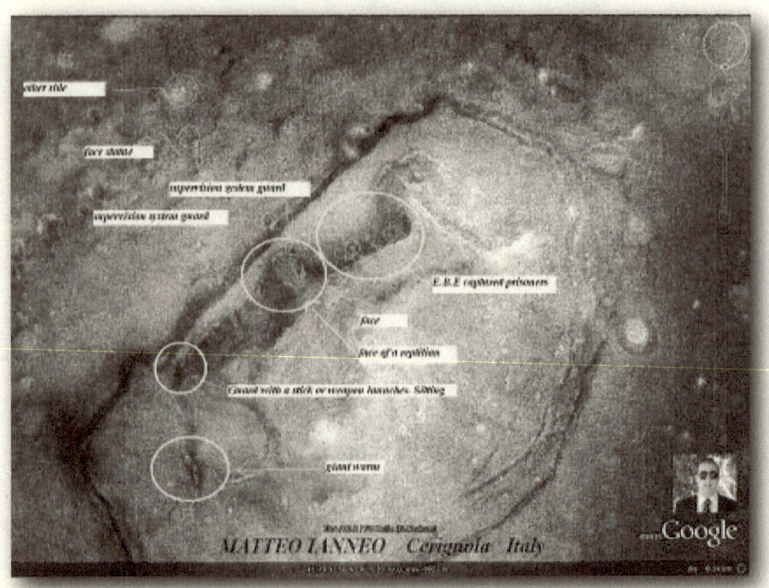

Image source: ESA/DLR/FU Berlin (G.Neukum)

At the centre you can see a shape similar to a reptile. To be precise, the head of a reptile. If you look to the bottom left, you can see a sort of giant worm, similar to a cockroach. Above that, someone is sitting on a kind of chair – perhaps mobile – holding a spear in his hands. This could be an artefact that emanates electric shocks, used to keep captured aliens under control.

But this is only my hypothesis.

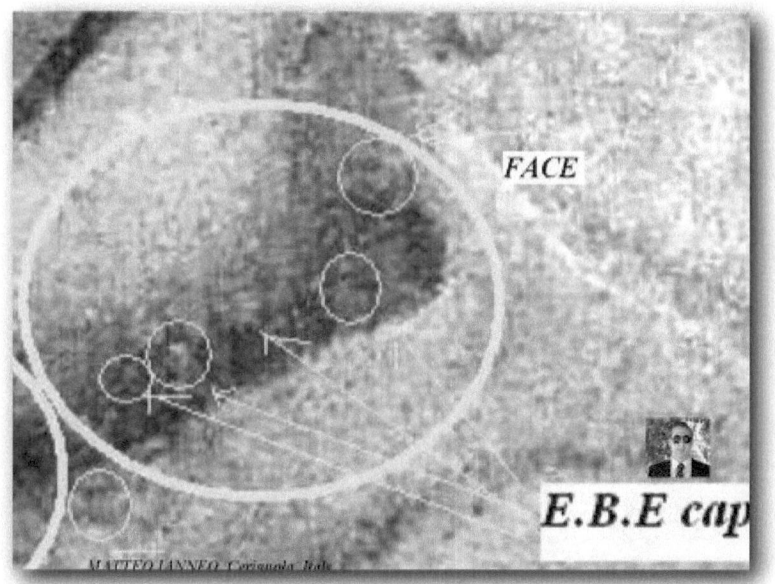

Image source: ESA/DLR/FU Berlin (G.Neukum)

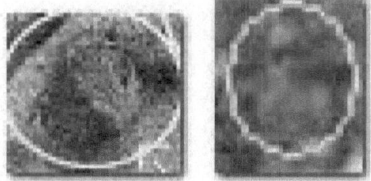

Here I have zoomed in on part of the image to highlight the creatures caught in the ditch and kept at bay by someone. At the top, if you look closely, there is an e.b.e., you can see her dark eyes.

Let's continue our exploration.

Another profile makes its appearance showing a face with a hat on its head.

The position on
Google Earth is as follows:
Latitude 22°38'11.69"N   Longitude 132°56'18.16"W

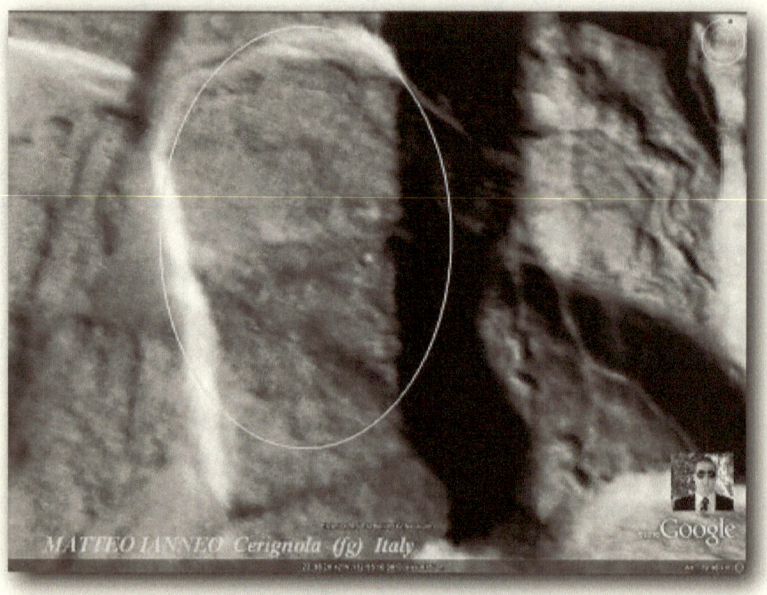

Image source: ESA/DLR/FU Berlin (G.Neukum)

A portrait profile, it is possible to see that it is equipped with a nose, mouth and hat, I'd say a Busby. Another element that was slowly filling my collection of faces and profiles.

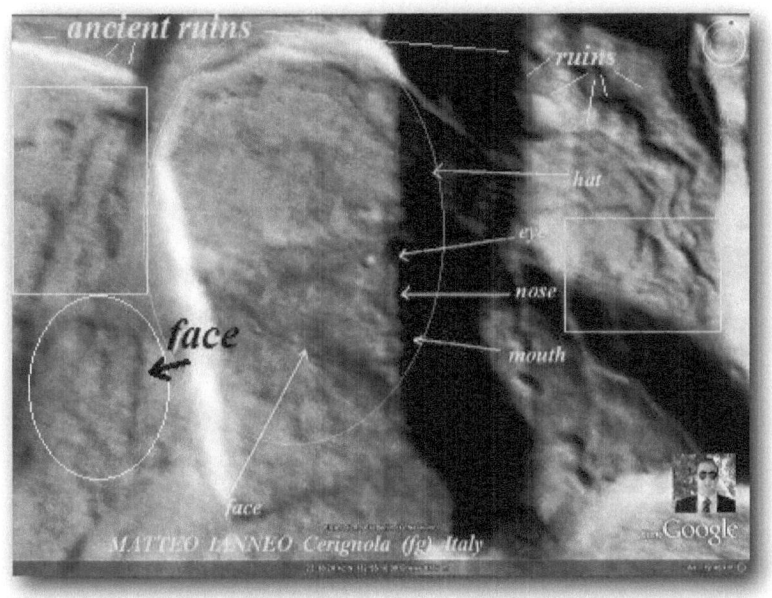

Image source: ESA/DLR/FU Berlin (G.Neukum)

Here in more detail, we can observe its features.

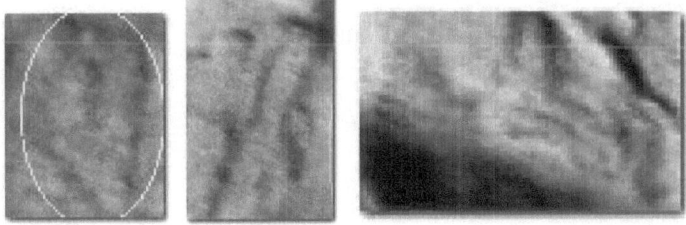

Finally, I found a good example of a lion carved into the rock.

The position on

Google Earth is as follows:

Latitude 41°11'30.11"N   Longitude 14°15'27.11"E

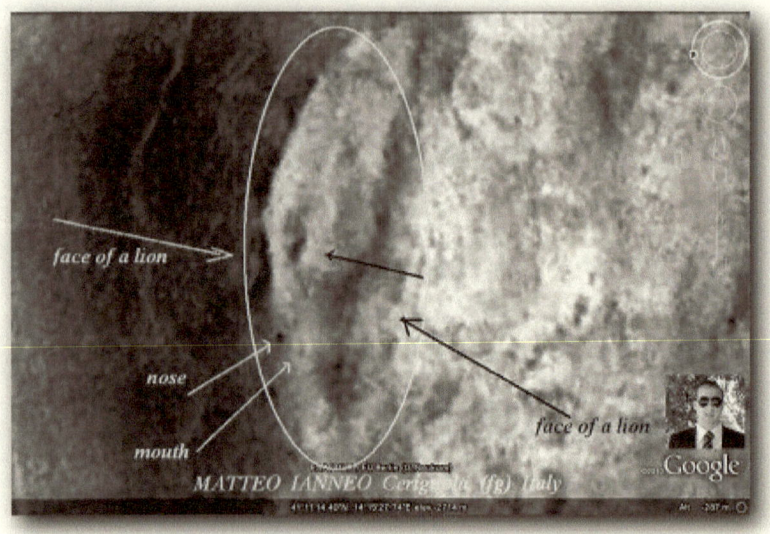

Image source: ESA/DLR/FU Berlin (G.Neukum)

Here it is carved into the mountain. Notice the head, a closed eye, pronounced nose and mouth. Nearby you can also see the remains of some ruins.

Image source: ESA/DLR/FU Berlin (G.Neukum)

Here it is in detail.

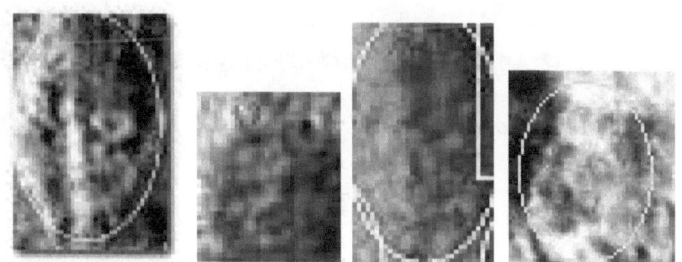

If you notice, there are several details of the monuments that belong to this civilisation. Usually, in fact, near to the faces we often find the remains of an ancient civilisation.

Indeed, we also find the kind of depictions that resemble totems here on Earth.

The position on

Google Earth is as follows:

Latitude 41° 9'4.60"N   Longitude 14°17'31.90"E

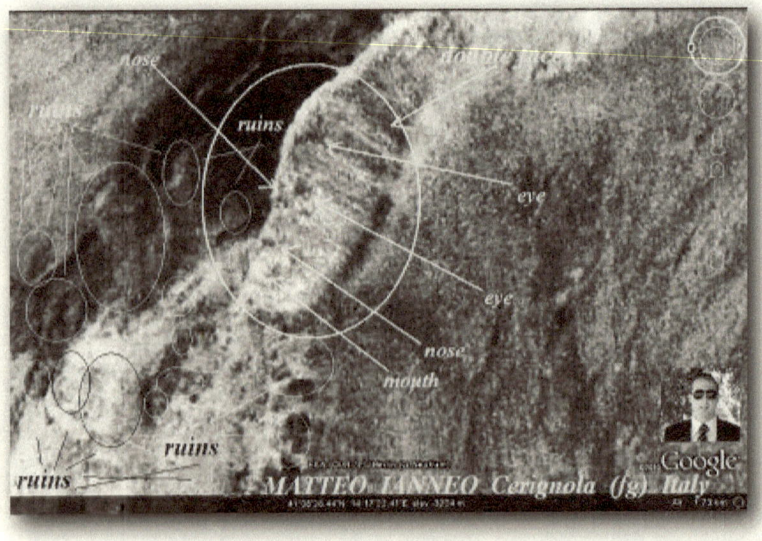

Image source: ESA/DLR/FU Berlin (G.Neukum)

Here is a face in the rock surrounded by ruins.

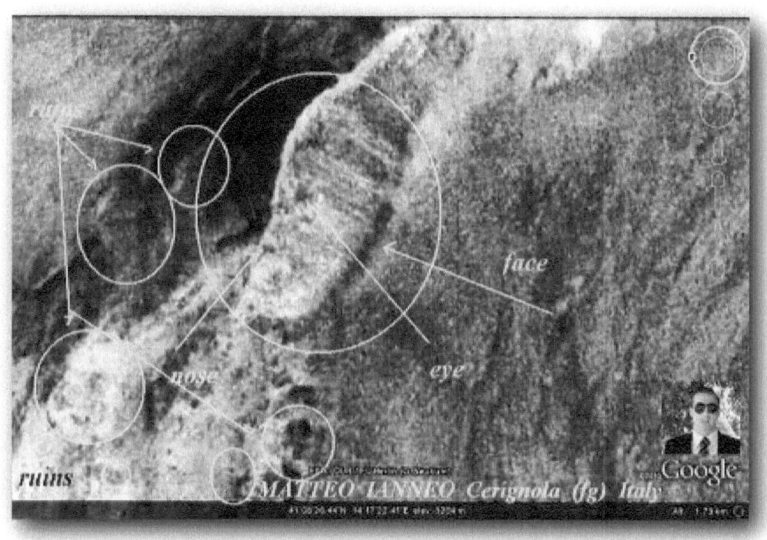

Image source: ESA/DLR/FU Berlin (G.Neukum)

Here you can see the ruins in more detail located in the area of the main subject.

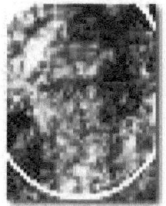

This element is, to say the least, amazing. Here is the face of a very old being.

A prophet.

The position on

Google Earth is as follows:

Latitude 42°00'31.63"N   Longitude 44°42'45.11"E

Image source: NASA / JPL / University of Arizona

This face is spectacular. It had to be the face of a prestigious being of the planet. Someone who wanted to leave their mark, dedicating this wonderful sculpture – a creature worthy of admiration. You can see the eye, the nose of a sorcerer, mouth and elongated chin – a characteristic feature.

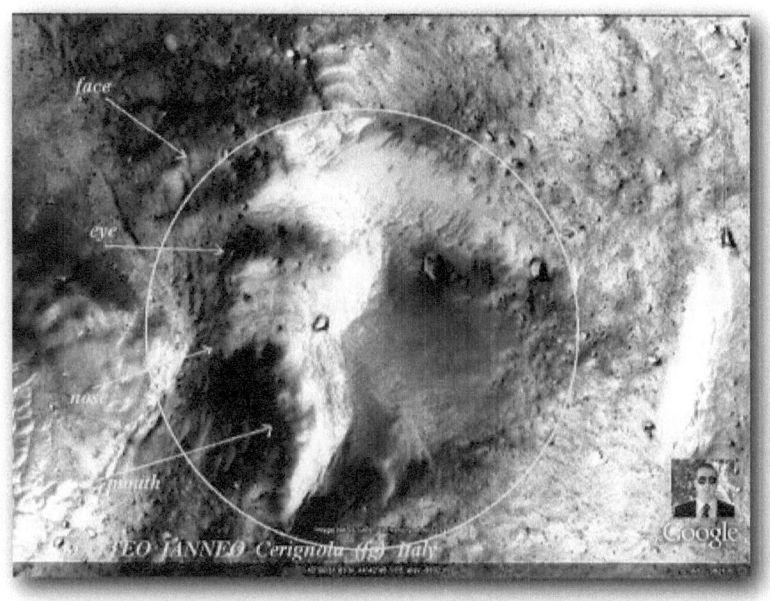

Image source: NASA / JPL / University of Arizona

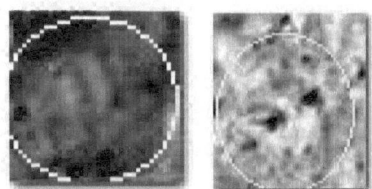

I didn't know if ghosts had ever come to Mars, but I found a ghost similar to the ones found in children's cartoons.

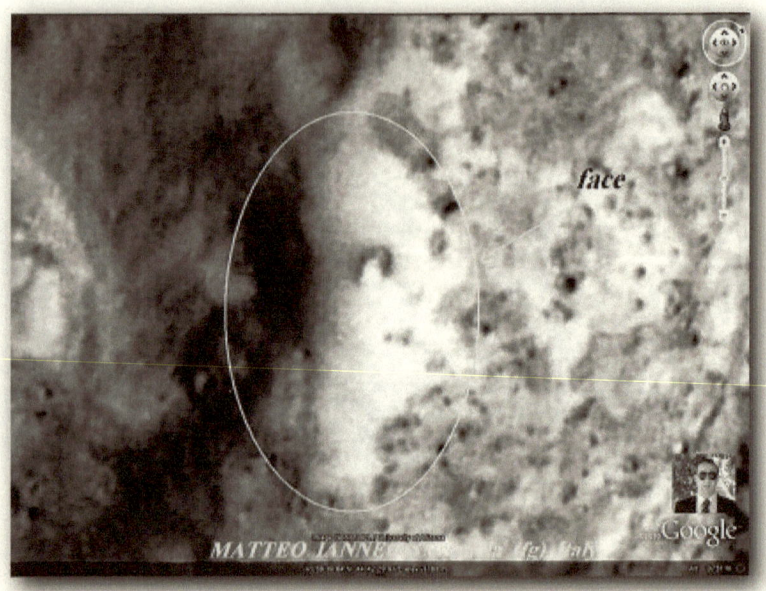

Image source: NASA / JPL / University of Arizona

Here it is, posing for us - smiling and happy to be seen. You can see the mouth, eyes, nose and its smile.

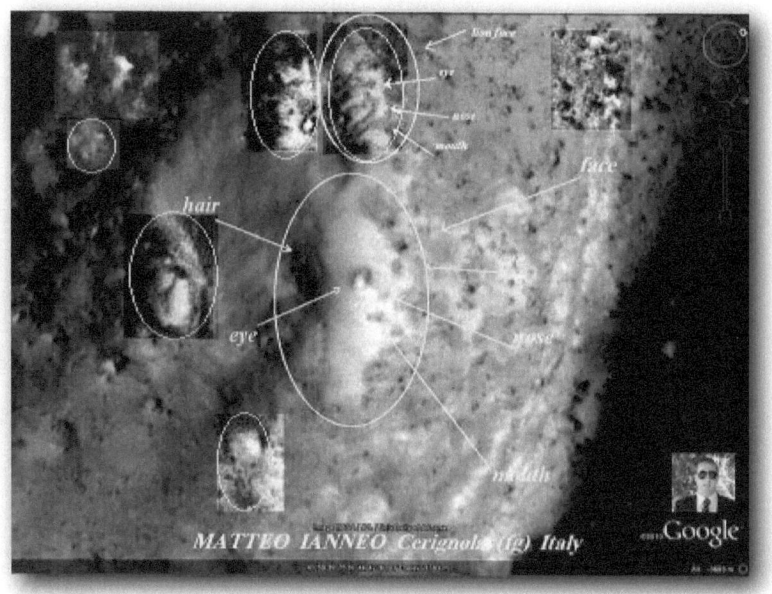

Image source: NASA / JPL / University of Arizona

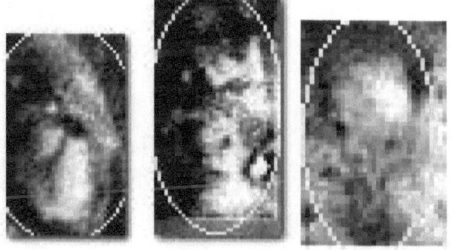

Above it, you can see the profile of the time-worn face of a being. So, as you see, the faces are never alone, they are always accompanied by something that makes them more true. And then... someone on the mountain shouts out "I am here too! I'm here, Can't you see me?".

Ah, there he is! An alpine soldier is calling us.

The position on

Google Earth is as follows:

Latitude 34°48'32.09"N   Longitude 65°37'34.63"E

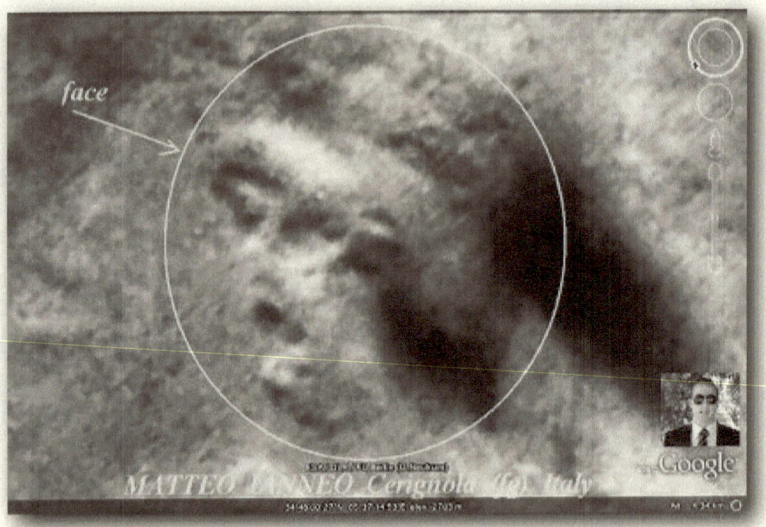

Image source: ESA/DLR/FU Berlin (G.Neukum)

The mountain soldier is screaming out to be seen and has a hat on his head, typical of areas where the temperature is very low. His mouth shows an entrance to access underground.

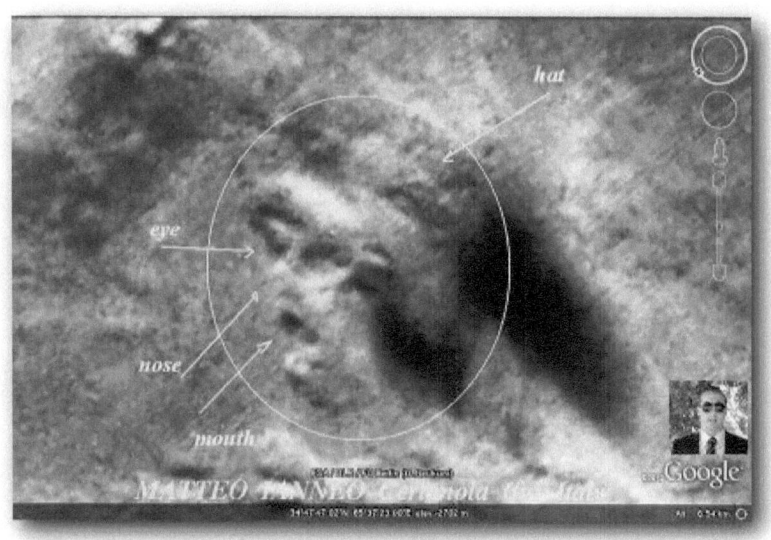

Image source: ESA/DLR/FU Berlin (G.Neukum)

Image source: ESA/DLR/FU Berlin (G.Neukum)

Other details.

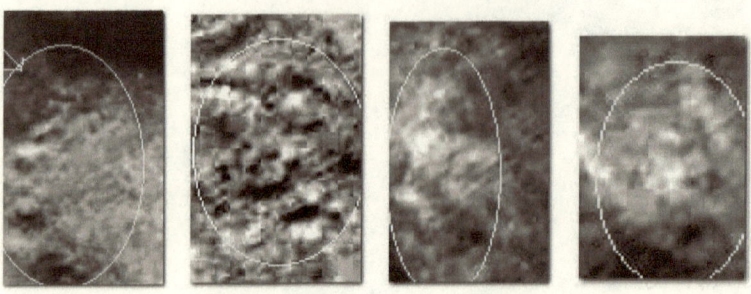

But he shouldn't shout so loudly because he could wake up the sleeping lady.

The position on

Google Earth is as follows:

Latitude 18°13'43.03"N   Longitude 44° 9'39.46"W

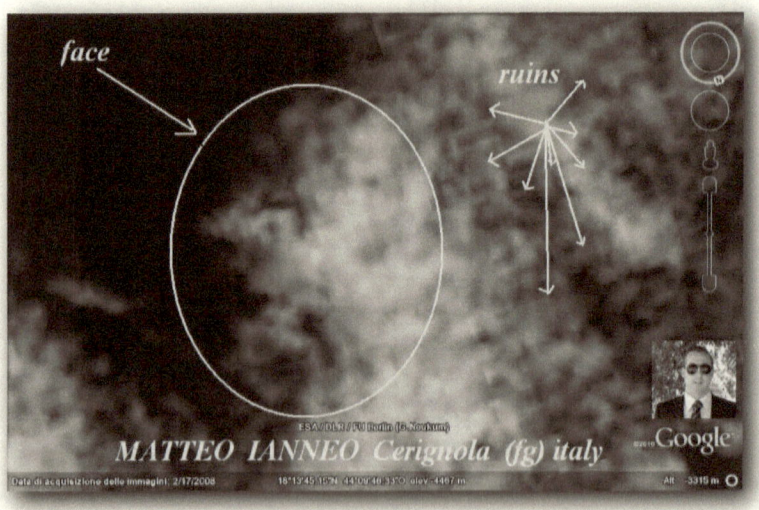

Image source: ESA/DLR/FU Berlin (G.Neukum)

Here is a beautiful sleeping goddess. She is waiting for her prince to wake her. You can see her eye, nose, mouth, and nearby, yet again ruins.

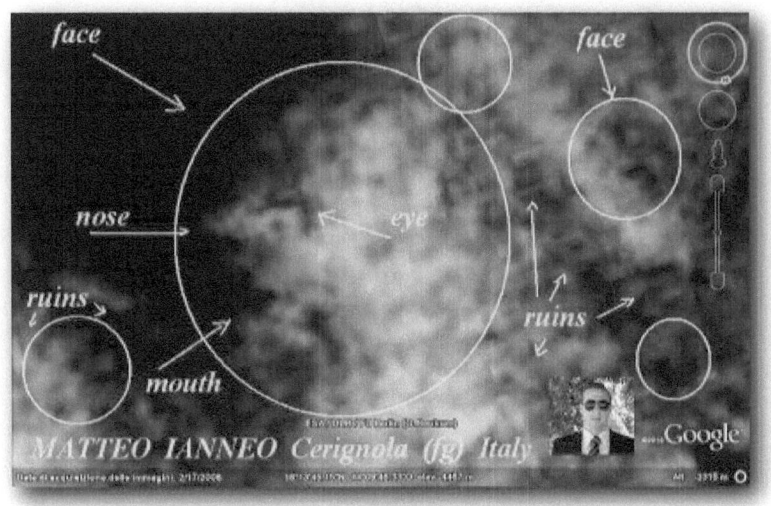

Image source: ESA/DLR/FU Berlin (G.Neukum)

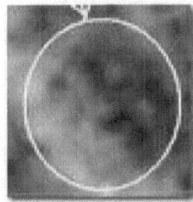

If you look to the top right, there is another feline and other details. We need a leader to stay calm. We need someone to take our defence.

A leader.

The position on
Google Earth is as follows:
Latitude 18°12'37.42"N   Longitude 44°11'17.41"W

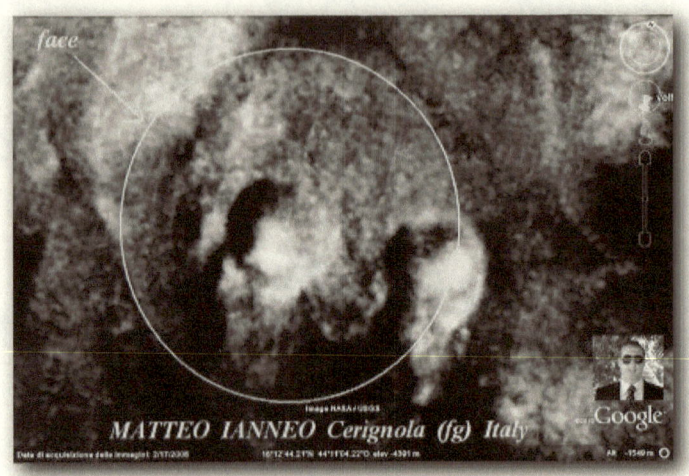

Image source: NASA / USGS

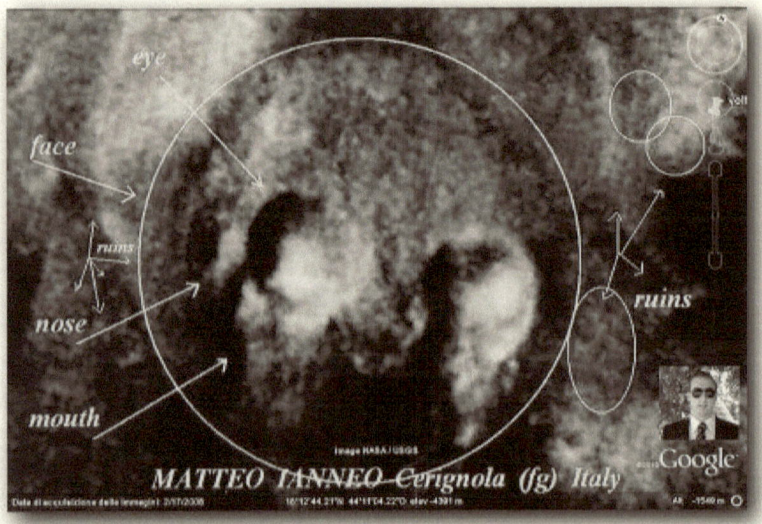

Image source: NASA / USGS

Here it is in detail, accompanied by the ruins.

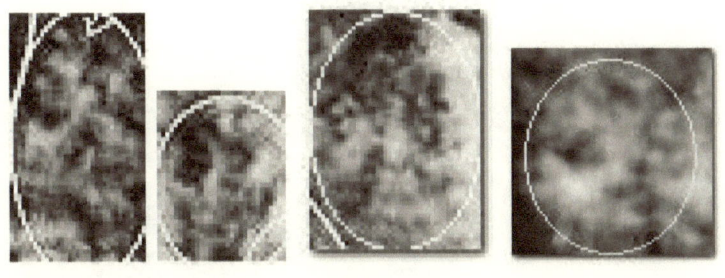

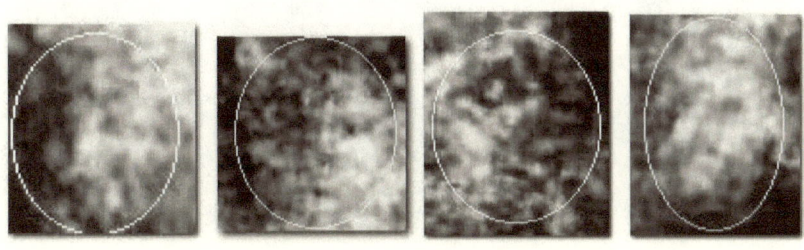

The next frame, I've entitled *The Lovers.*

The position on

Google Earth is as follows:

Latitude 18°12'59.33"N   Longitude 44°04'09.89"W

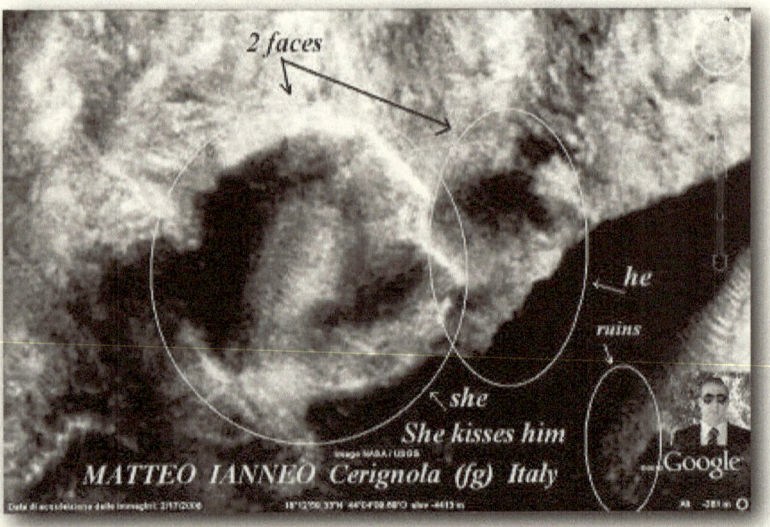

Image source: NASA / USGS

She is kissing him. Perhaps in the history of Mars there were two inseparable gods. So, in this photo, we see a woman's profile kissing a man's profile. Nearby we see the usual ruins of an ancient civilisation.

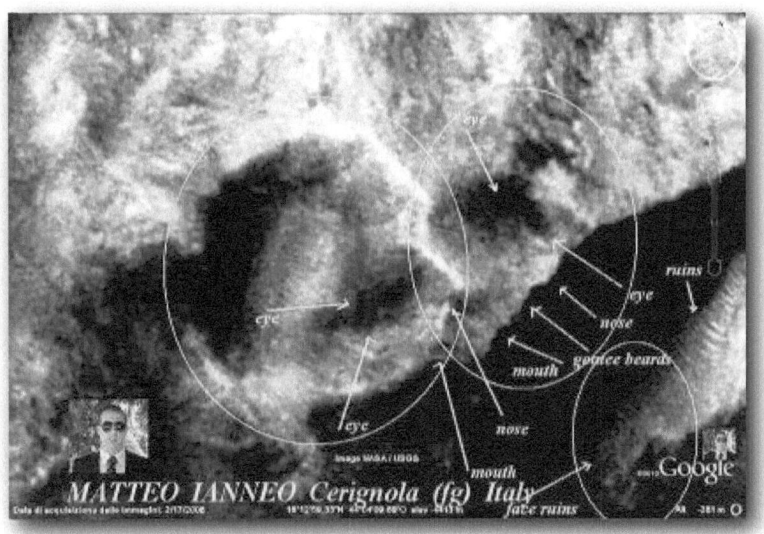

Image source: NASA / USGS

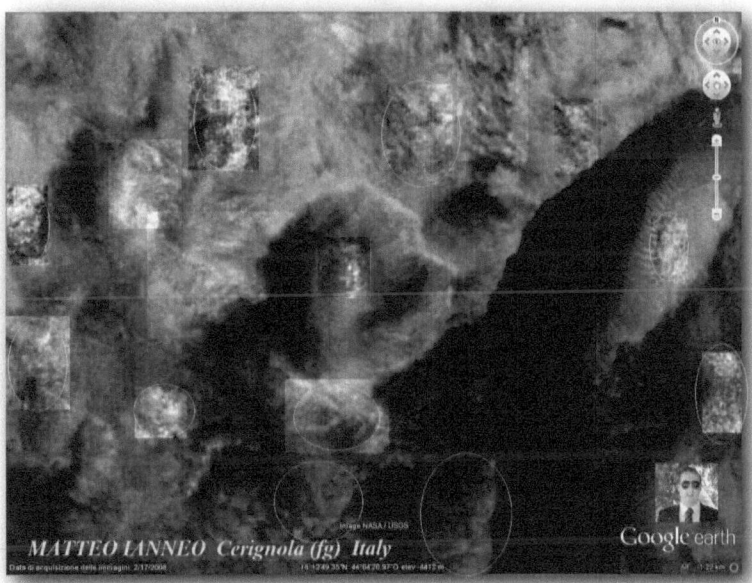

Image source: NASA / USGS

Here are the details.

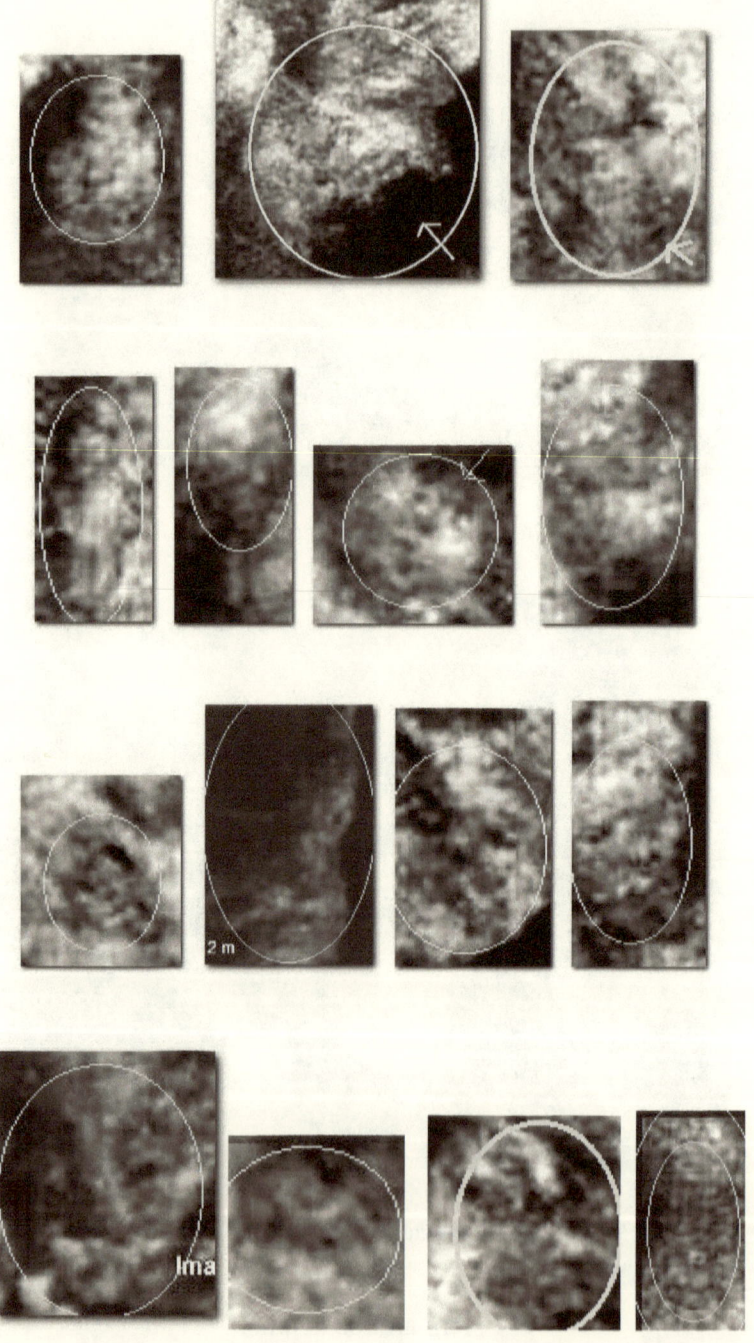

In the next photo we can see a funny face again.

The position on
Google Earth is as follows:
Latitude 40°45'56.79"N   Longitude 14°41'57.94"E

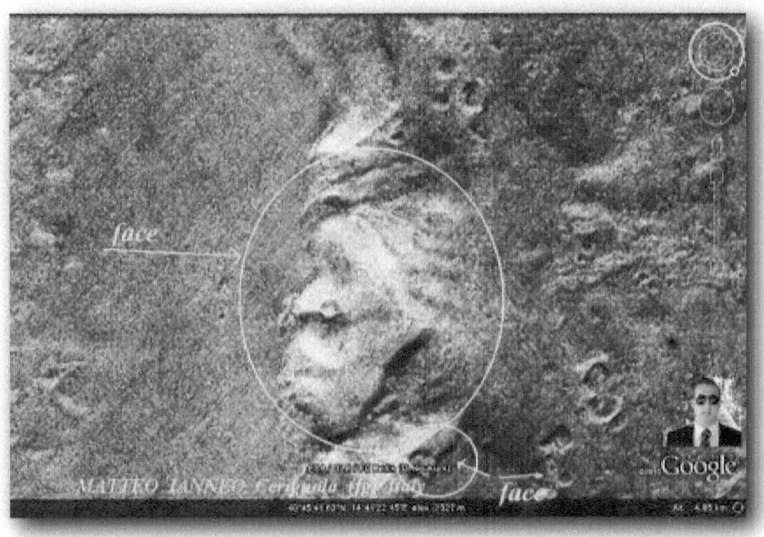

Image source: ESA/DLR/FU Berlin (G.Neukum)

A face with a very obvious jaw, also eroded by time. We can see the nose, but the eyes and mouth are not very visible.

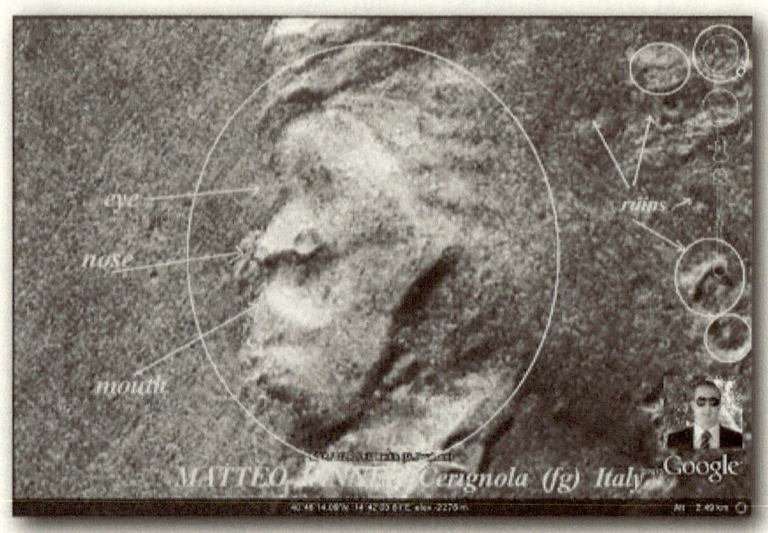

Image source: ESA/DLR/FU Berlin (G.Neukum)

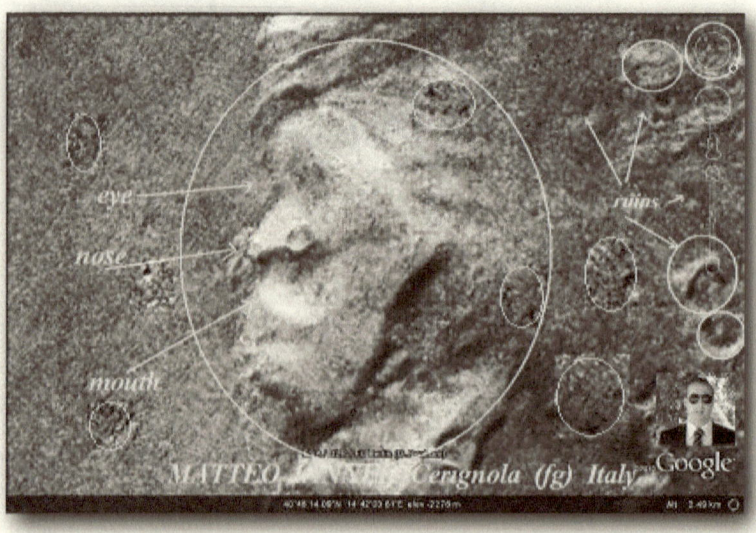

Image source: ESA/DLR/FU Berlin (G.Neukum)

Details of the ruins.

If you have a good eye, you will see some remains of ancient ruins to the right of the image.

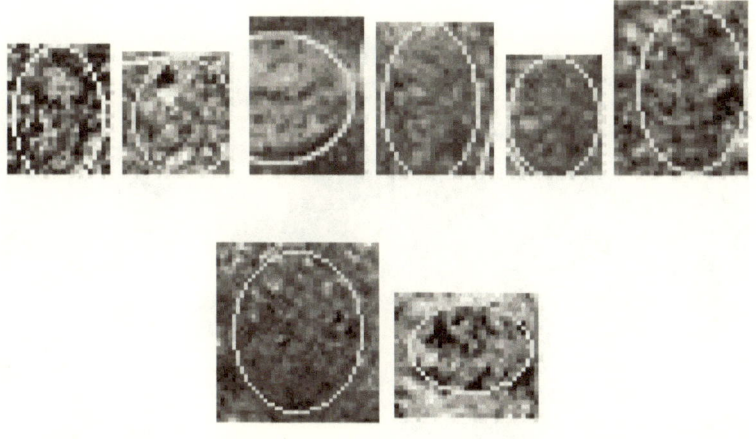

When I was little, I had a dog called Pimpi. One day he ran away from home and I never found him again. My tears, in those days, were those of despair. After many years, I found him again here, on the Red Planet. He seems afraid of something, perhaps for having forgotten the way back.

The position on

Google Earth is as follows:

Latitude 42°20'1.26"N   Longitude 0°22'42.48"W

Image source: NASA / USGS

Pimpi I'm here, I'm looking at you, but you can't see me.

In this crater there is a dog's head. You can clearly see its almond-shaped eyes, his muzzle, and his ears. Too clearly to be another effect of nature. Anyway, let's put it in the realm of fantasy.

Image source: NASA / USGS

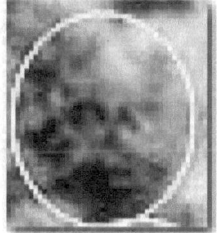

Look to the top centre, you can see the remains of ruins, an almost skeletal face with eyes, nose and mouth. Then to the left of the photo other details. Hi, Pimpi!

After all of these anomalies is it likely to find something special: the remains of a sphynx.

The position on

Google Earth is as follows:

Latitude 38°16'48.30"N   Longitude 13° 3'39.79"W

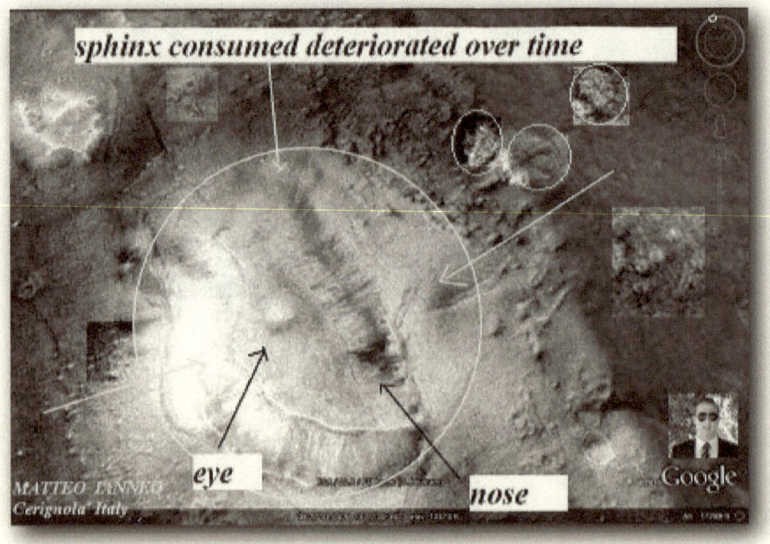

Image source: ESA/DLR/FU Berlin (G.Neukum)

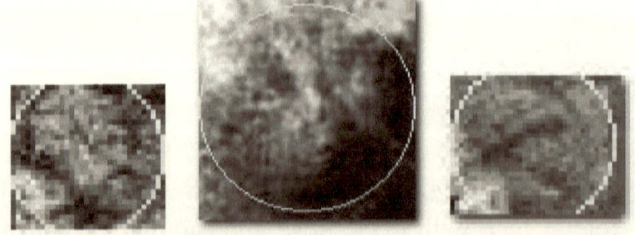

Here it is, almost half of a sphynx. The other part has eroded away over time, but I wanted to show it to you because it retains the following very important details: a closed eye and the nose of a lion. Surrounding it there are further details.

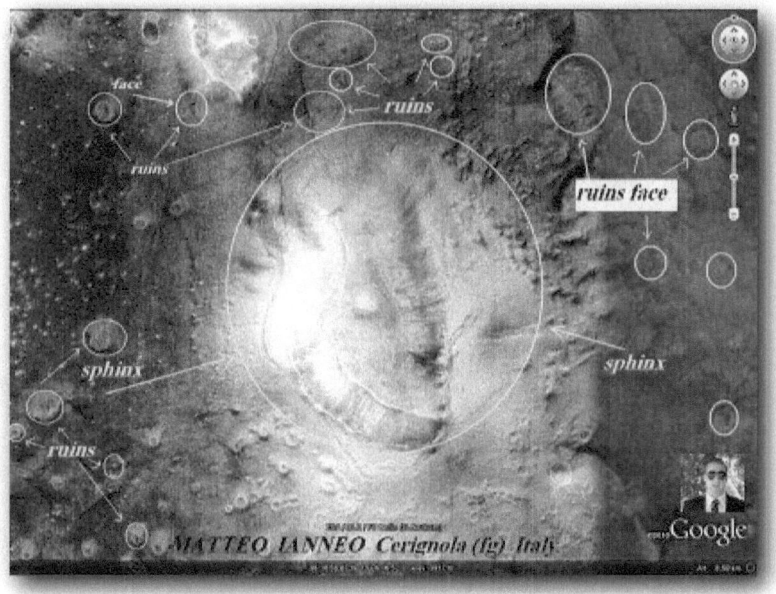

Image source: ESA/DLR/FU Berlin (G.Neukum)

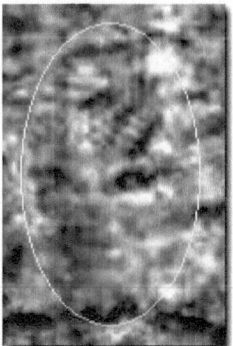

Look to the top right, you can see another monumental face and other details. To the left, other ruins eroded over time.

I isolated the monumental profile and highlighted it by zooming in. Here is the profile.

The position on

Google Earth is as follows:

Latitude 38°20'54.35"N   Longitude 12°59'19.25"W

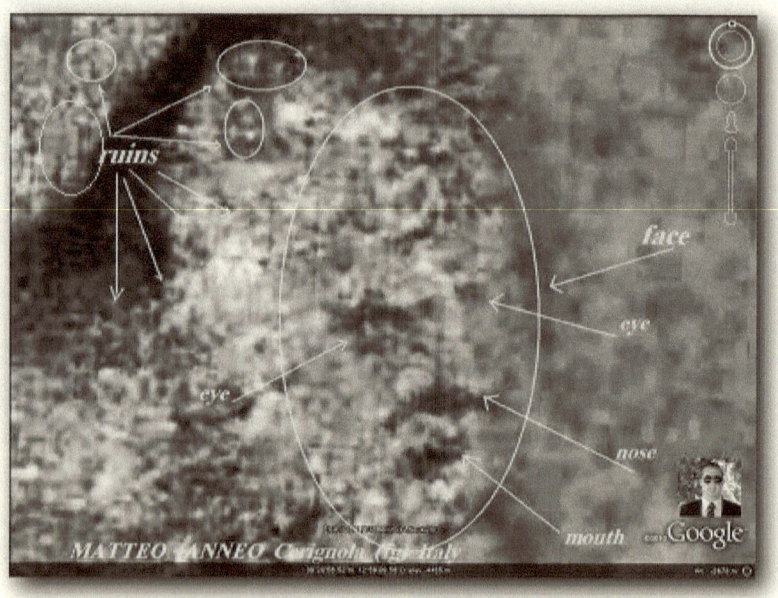

You can see the pupil of the left eye, the nose and mouth. To the top, small, you can see the ruins of an ancient civilisation.

It would need strong magnification to see the ancient city in detail.

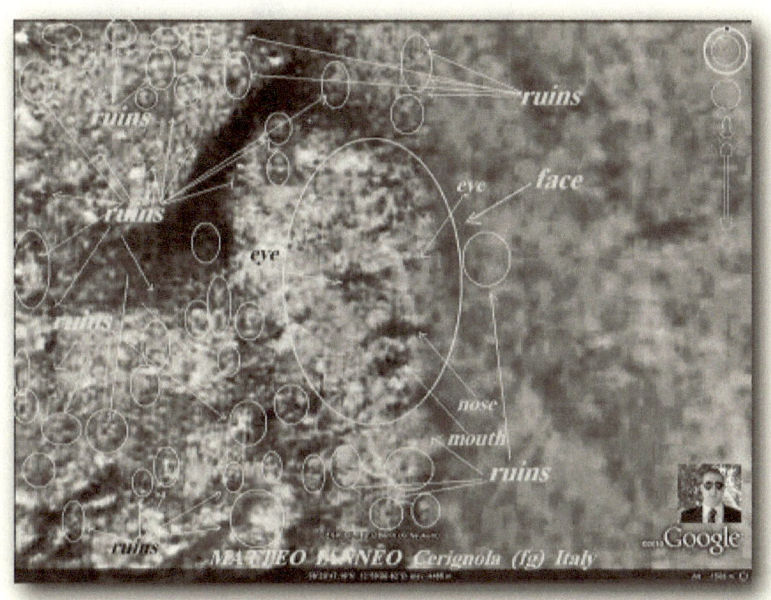

Image source: ESA/DLR/FU Berlin (G.Neukum)

Image source: ESA/DLR/FU Berlin (G.Neukum)

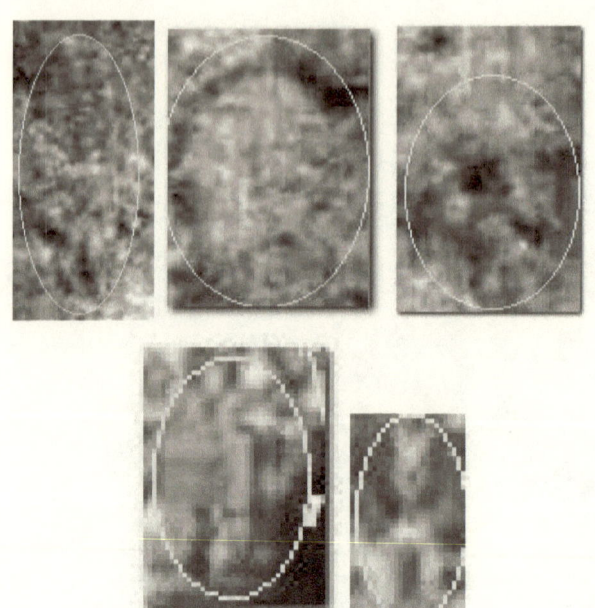

The position on
Google Earth is as follows:
Latitude 38°23'18.18"N Longitude 13° 1'3.59"W

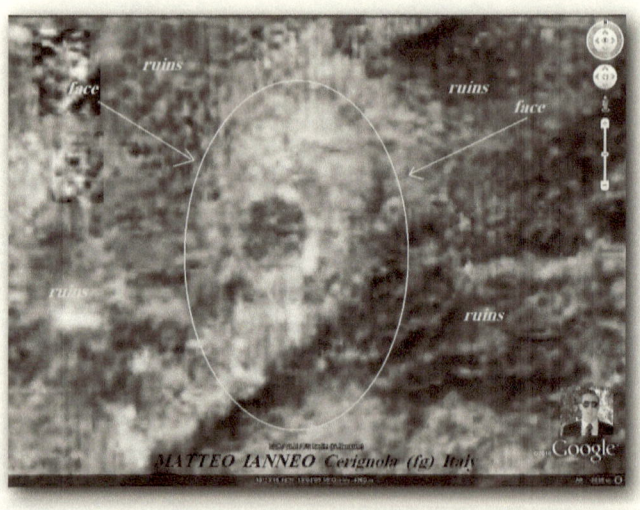

Image source: ESA/DLR/FU Berlin (G.Neukum)

A particular face.

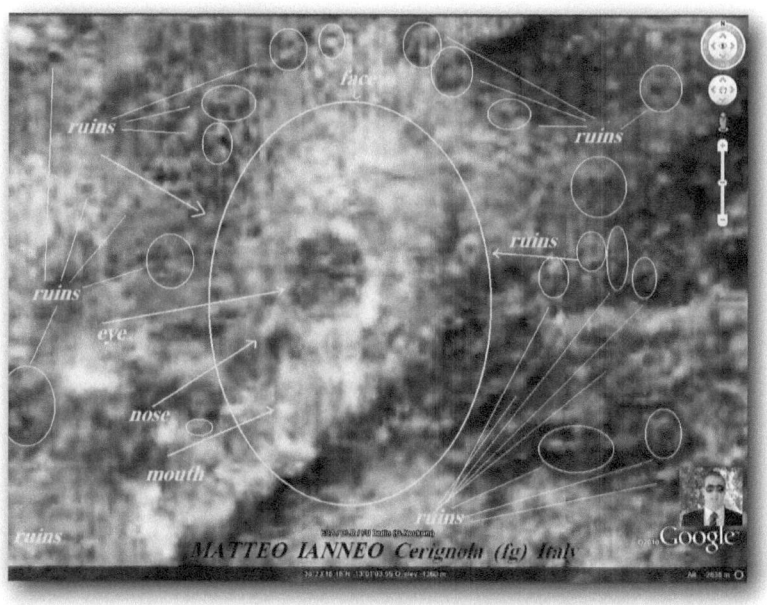

Image source: ESA/DLR/FU Berlin (G.Neukum)

We can see a small ear, nose, and very big eye.
Accompanied by ancient ruins.

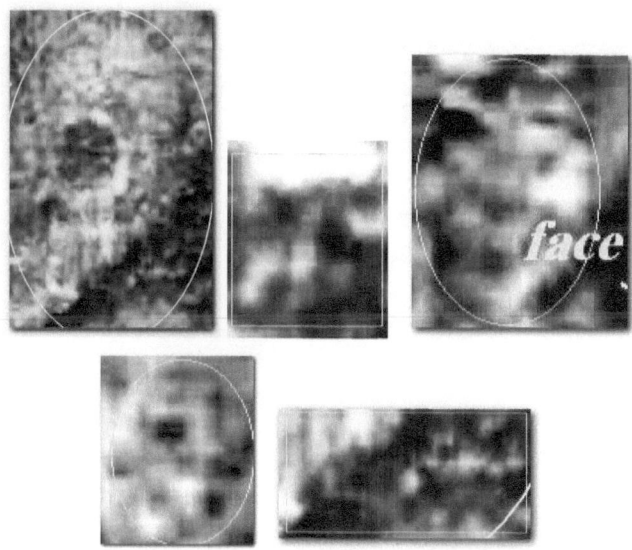

Image source: ESA/DLR/FU Berlin (G.Neukum)
NASA / USGS

In this photo we can see a winged creature; it looks like a bird of prey.

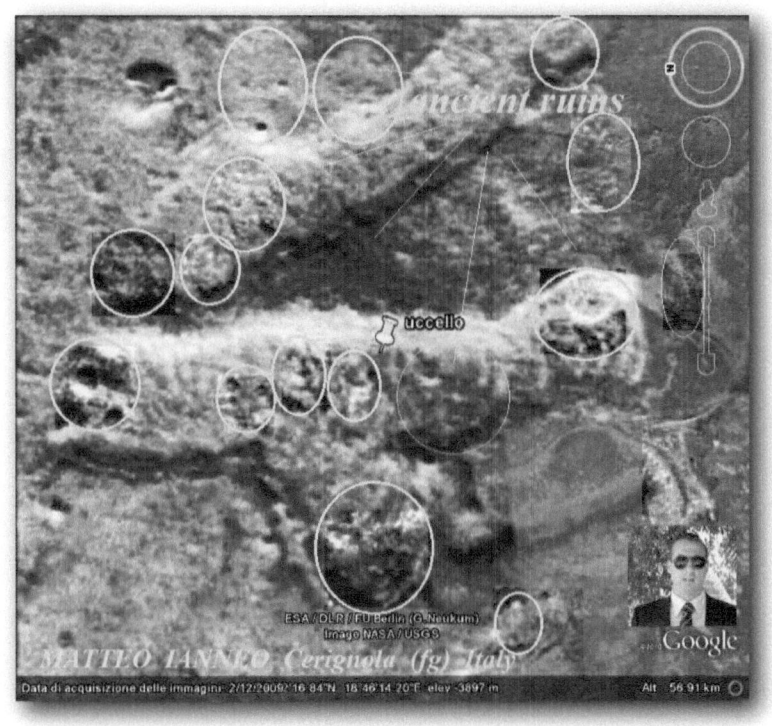

Image source: ESA/DLR/FU Berlin (G.Neukum)
NASA / USGS

Here are highlighted details.

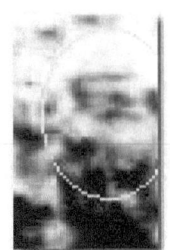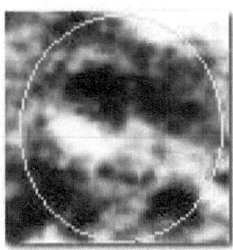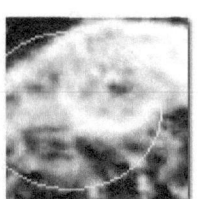

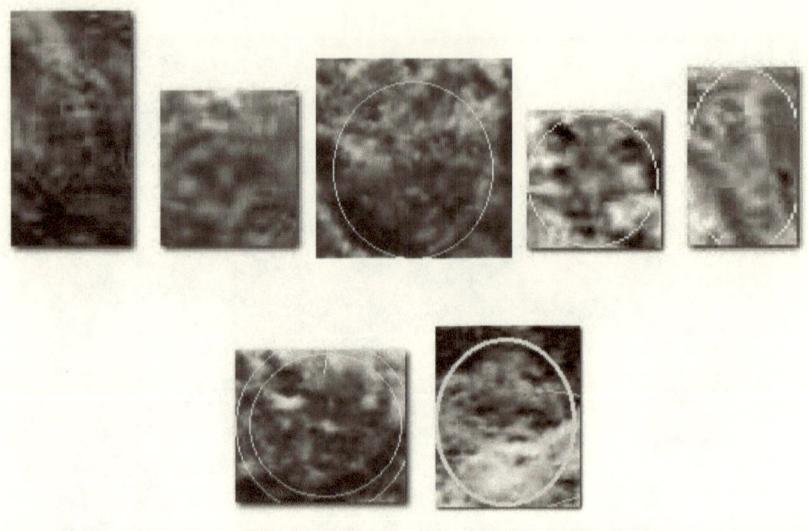

Remember that every face examined by me is not the result of erosion caused by natural elements, but hidden wonders created by people who dominated this extraordinary planet. If we were there, in this place we are looking at, we would find a huge expanse of kilometres where, walking, we would be surrounded by statues, figurines, monuments and so on - in short, a truly great city. We have reached the end of my first book on the study of the anomalies of the planet Mars. Before leaving you, I would like you to look at this last discovery that made me incredibly happy: the face of the Holy Shroud.

The position on

Google Earth is as follows:

Latitude 38°20'49.26"N Longitude 13° 2'5.56"W

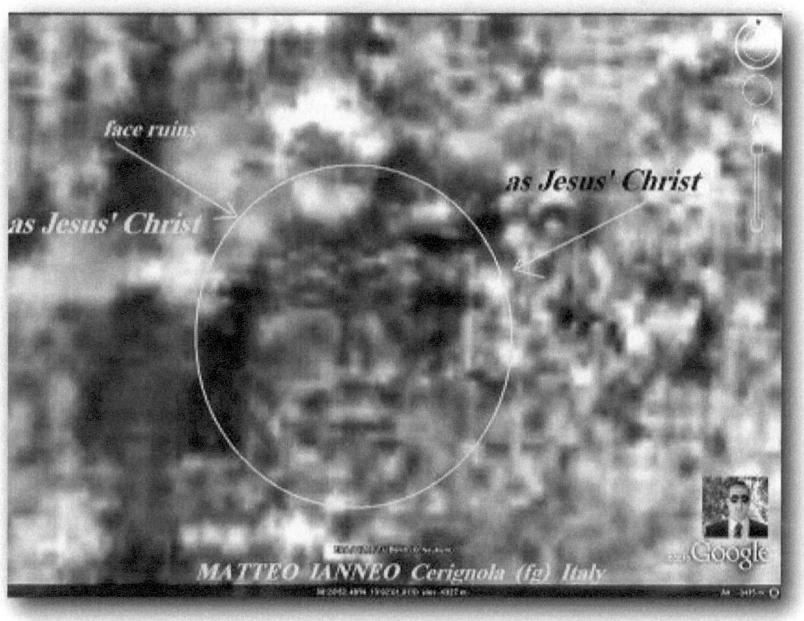

Image source: ESA/DLR/FU Berlin (G.Neukum)

It was the first thing that came to mind when I made the discovery. Comparing this anomaly with the Holy Shroud, I realised that they were exactly the same.

The nose, mouth and eyes are identical. Do you think there could be pixellation? Yes, there could be, but if there wasn't, the image would be much clearer.

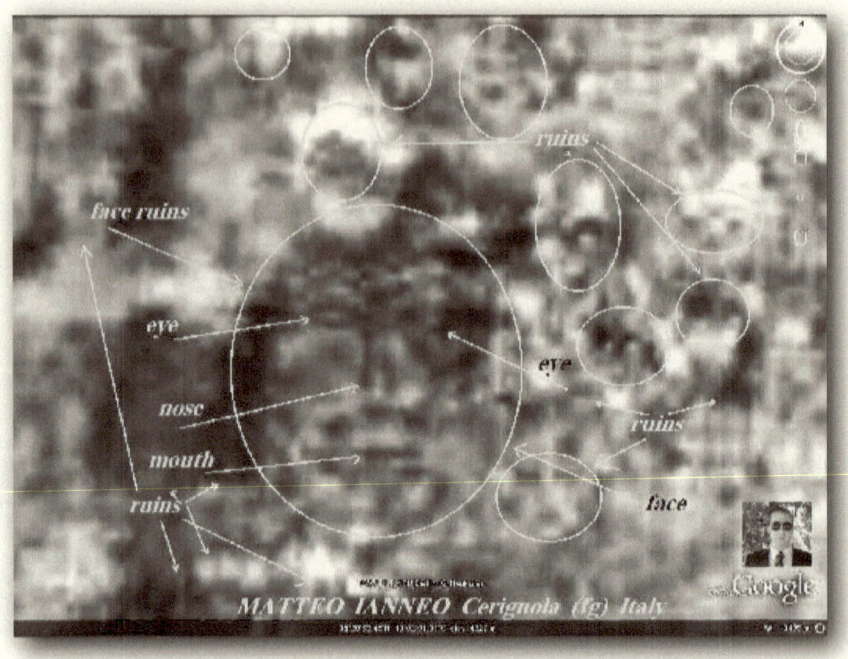

Image source: ESA/DLR/FU Berlin (G.Neukum)

Details.

To the bottom left you can see – even if not very obvious – ruined columns which certainly supported a temple in this city.

Conclusion

I hope this voyage has been to your liking. I wanted to present this book to give a small contribution to those, like me, who have a vision of history different from the one we know. I conducted these studies alone and with great effort. I had to probe the planet in detail with my little instruments in order at least to obtain details and elements that could be seen by everyone. There are many things that I had to discard because they were not easy to comprehend. I have kept back other elements that I will publish in the next volume.

I thank, especially, the Google Earth *tool* for offering this instrument – thanks to it, I was able to conduct my studies and obtain the results reported in this book.

I thank NASA and ESA for the satellite images produced by them as a source of information for my studies.

ESA/DLR/FU Berlin (G. Neukum), NASA/USGS.

My thanks to all those who had faith in me and supported me psychologically on this difficult and tormented voyage full of positive feedback, but above all, negative. My studies started in 2009, the year in which I identified most of the elements presented, and continue today. Many experts in the subject have done nothing but belittle the elementary studies of an unqualified person in the field who does not have a basic preparation in the subject. I thank them all the same.

My greetings to you all. See you for the next wonderful voyage.

www.ingramcontent.com/pod-product-compliance
Lightning Source LLC
Chambersburg PA
CBHW032016170526
45157CB00002B/727